AF540802

# TEXT BOOK OF CYTOGENETICS

# TEXT BOOK OF CYTOGENETICS

*By*

**Dr. Rajiv Tyagi**

*Department of Zoology*

*M.M. College*

*Modi Nagar (U.P.)*

*(India)*

*Published by:*

**DISCOVERY PUBLISHING HOUSE**
4383/4B, Ansari Road, Darya Ganj
New Delhi-110 002 (India)
*Phone* : +91-11-23279245; 23253475; 43596065
*Mobile* : +91 9811179893 / +91 9871656464
*E-mail* : discoverybooksindia@gmail.com
orderdphbooks@gmail.com
namitwasan9@gmail.com
*web* : www.discoverypublishinggroup.com

***First Published:*** **2009**

***Reprinted:*** **2026**

**ISBN: 978-81-8356-496-0**

**Text Book of Cytogenetics**



*Printed at:*
Infinity Imaging Systems
Delhi

# Preface

Cytogenetics is the study of chromosomes and the related disease states caused by numerical and structural chromosome abnormalities. A variety of cell or tissue types can be used to perform these studies.

Cytogenetics is the study of abnormalities in chromosomes. Chromosomes are obtained from dividing cells. The collection of chromosomes from a single cell is known as a metaphase spread.

Bone marrow is one of the few tissues in the body where spontaneous cell division occurs throughout life. The marrow is actively producing the constituent cells of the blood, red and white cells and platelets, which have life spans of varying lengths. Bone marrow can be cultured by adding a small amount of bone marrow sample to a culture tube containing a nutrient rich medium; metaphase spreads can be obtained from these cultures within a few hours.

Solid tissue samples, however, need to be incubated in culture flasks in nutrient rich medium for up to two weeks to obtain sufficient dividing cells for cytogenetic investigations to be carried out.

After a bone marrow aspirate or solid tissue sample has been taken from a patient, it is placed in special culture fluid and transported to the cytogenetics laboratory for processing. Cytogeneticists are trained to examine the chromosomes obtained from the sample and detect disease-related abnormalities.

The different cytogenetic changes seen in malignancy have been catalogued from data obtained during large collaborative studies and clinical trials. Cytogenetic results, in combination with other data are classified as giving good, standard or poor risk chance of survival to patients. These "risk groups" now provide clinicians with a prognostic indicator of likely patient outcome, enabling structured treatment for any individual within a common framework of tried and tested treatments.

The results of cytogenetic investigations are also used to identify new recurrent abnormalities that may have prognostic significance for patients, and to investigate the aetiology and mechanisms of malignancy.

**—Author**

# Contents

# 1

# Introduction

Cytogenetics is the study of chromosomes and the related disease states caused by numerical and structural chromosome abnormalities. A variety of cell or tissue types can be used to perform these studies.

## Introduction to Cytogenetic Studies in Malignancy

Cytogenetics is the study of abnormalities in chromosomes. Chromosomes are obtained from dividing cells. The collection of chromosomes from a single cell is known as a metaphase spread.

Bone marrow is one of the few tissues in the body where spontaneous cell division occurs throughout life. The marrow is actively producing the constituent cells of the blood, red and white cells and platelets, which have life spans of varying lengths. Bone marrow can be cultured by adding a small amount of bone marrow sample to a culture tube containing a nutrient rich medium; metaphase spreads can be obtained from these cultures within a few hours.

Solid tissue samples, however, need to be incubated in culture flasks in nutrient rich medium for up to two weeks to obtain sufficient dividing cells for cytogenetic investigations to be carried out.

After a bone marrow aspirate or solid tissue sample has been taken from a patient, it is placed in special culture fluid and transported to the cytogenetics laboratory for processing. Cytogeneticists are trained to examine the chromosomes obtained from the sample and detect disease-related abnormalities.

## Clinical Applications of the Results of Cytogenetic Testing

The different cytogenetic changes seen in malignancy have been catalogued from data obtained during large collaborative studies and clinical trials. Cytogenetic results, in combination with other data are classified as giving good, standard or poor risk chance of survival to patients. These "risk groups" now provide clinicians with a prognostic indicator of likely patient outcome, enabling structured treatment for any individual within a common framework of tried and tested treatments.

The results of cytogenetic investigations are also used to identify new recurrent abnormalities that may have prognostic significance for patients, and to investigate the aetiology and mechanisms of malignancy.

## Cell Cultures and Slide Preparation

After the incubation period, cells that are in the process of cell division (mitosis) are, by the addition of a chemical to the cultures, stopped at the metaphase stage of the cell cycle. This is the stage in the cell division cycle where the chromosomes condense and will appear as distinct elements which can be seen using a microscope.

The cells are treated with hypotonic solutions that swell the nuclei and allow the chromosomes to float freely within the cell membrane. The cells are then fixed with an acetic acid and methanol mixture. A drop of this mixture containing the cells is dropped onto a microscope slide.

When the drop of cell suspension hits the slide, the cell nuclear membranes break and the chromosomes inside the cell nucleus spread out and can be viewed as individual elements.

The slides are dried and treated with special staining techniques to produce the characteristic stripy appearance (banding patterns) seen in chromosome preparations.

## Microscopic Examination of Chromosome Preparations

The slides are examined using a light microscope and scanned to find metaphase spreads which appear as discreet bundles of chromatin.

They are magnified up to 1200 times and the 46 individual chromosomes are matched into 23 pairs by means of their own distinct banding pattern. These pairs are arranged by size and numbered from 1, the largest, to 22, with the final pair of sex chomosomes being called 'XX' for a female and 'XY' for a male. A trained cytogeneticist can identify any discrepancies in the number or structure of the chromosomes.

## Automated Chromosome Analysis

To assist with the identification of the chromosomes automated analysis systems have been developed which provide an on-screen image of the chromosomes using a camera mounted on the microscope. These systems allow manipulation of the chromosomes on screen to form the karyogram in which the chromosomes are paired and any inconsistencies within the pairs can be seen.

Abnormalities of number or structure which are found in the chromosomes could be an indicator of disease and may aid diagnosis and prognosis of leukaemia.

## Chromosome Abnormalities in Malignancy

Not all patients with leukaemia or other malignancies have chromosome abnormalities, but where abnormalities are detected they can be a useful indicator, in parallel with other studies such as morphology (bone marrow structure), immunophenotyping (antibody type) and histology (tissue structure) in diagnosing the type of leukaemia or malignancy present.

The regions in which abnormalities occur have been shown to be associated with oncogenes (cancer causing genes) or regulatory genes, the disruption of which are the most likely cause of the leukaemic process, which is the uncontrolled proliferation of abnormal white blood cells.

The first consistent chromosome abnormality found in leukaemia was described in 1960 when workers in Philadelphia discovered that, in cases of chronic myeloid leukaemia (CML), one of the smallest chromosomes, no. 22, appeared to be smaller (deleted). Subsequently this was shown to be a balanced exchange of material (translocation) between chromosomes 9 and 22.

The rearrangement has subsequently been seen in cases of acute lymphoblastic leukaemia (ALL) and acute myeloid leukaemia (AML).

For many years this remained the only chromosome abnormality to be used in the diagnosis of a specific leukaemia. Though other abnormalities were seen in cases of leukaemia, it was only in the early 1970s with the advent of the special banding and staining techniques that many other consistently occurring abnormalities were detected in leukaemias of different kinds.

Other common recurrent abnormalities detected subsequently include: a translocation between chromosomes 15 and 17 't(15;17)' in acute promyelocytic leukaemia.

Additional copies of whole chromosomes are also seen, for example in some cases of ALL, where extra copies of chromosomes are gained in a characteristic manner. This raises the number of chromosomes in an abnormal cell from the normal (diploid) number of 46 chromosomes to an abnormal (high hyperdiploid) number, which may be as many as 60 chromosomes.

Recurrent abnormalities are also detected in cases of sold tumour; however obtaining metaphase spreads from solid tumour samples may be more difficult than bone marrow samples. An example of a recurrent abnormality in solid tumour is the translocation involving chromosomes 2 and 13 't(2;13)' associated with alveolar rhabdomyosarcoma.

There are many different types of leukaemia and other malignancies with consistent chromosomal changes which are seen much less frequently and more than one type of abnormality can be seen in a single case of leukaemia or other malignancy.

The elucidation of the exact abnormalities present in complex cases needs a highly-trained analyst and may require the use of complementary techniques.

## Comparative Genomic Hybridisation and Microarray Technology

In the 1990s advances in computer technology meant that FISH techniques could be used to detect relative gains and losses of chromosomal material by hybridising patient metaphases with normal ones. This is known as comparative genomic hybridisation (CGH).

This technique was further advanced by replacing the metaphase spreads with thousands of very small lengths of normal DNA in what is known as a microarray. The normal DNA sections are placed on a single slide and hybridised with fluorescently-labelled patient sample. The result of the hybridisation is read by a laser scanner. The information gained from the microarray hybridisation is fed into a computer programme and can give very accurate information about any genetic abnormalities that may present in the patient sample.

Genetics is the area of biological study concerned with heredity and with the variations between organisms that result from it. It demands an understanding of numerous terms, such as DNA (deoxyribonucleic acid), a molecule in all cells that contains blueprints for genetic inheritance; genes, units of information about particular heritable traits, which are made from DNA; and chromosomes, DNA-containing bodies, located in the cells of most living things, that hold most of the organism's genes. The vocabulary of genetics goes far beyond these three terms, as we shall see, but these are the core concepts. Among the areas in which genetics is applied is forensic science, or the application of science to matters of law—specifically, through "DNA fingerprinting," whereby samples of skin, blood, semen, and other materials can be used to prove or disprove a suspect's innocence. Another fascinating application of genetics is the Human Genome Project, an effort whose goals include the location and identification of every gene in the human body.

## Genetics and Heredity

Genetics and heredity, the subject of another essay in this book, are closely related ideas. Whereas heredity is the transmission of genetic characteristics from ancestor to descendant through the genes, genetics is concerned with hereditary traits passed down from one generation to the next. It is very hard, if not impossible, to separate the two concepts completely, yet the entire body of knowledge encompassed by these topics is so large and so complex that it is best to separate them as much as possible. For this reason, the Heredity essay is concerned with such issues as how traits are passed on and why they appear in a particular generation but not another. That essay addresses the topics of alleles, dominant and recessive genes, and so on. It also briefly discusses the history of studies in areas that encompass genetics, heredity, and the mechanics thereof. In general, the Heredity essay is concerned with the larger patterns of inheritance over the generations, while the present one examines inheritance at a level smaller than the microscopic—that is, from the molecular or biochemical level.

## Somatic and Germ Cells

Heredity begins with the cell, the smallest basic unit of all life. The information for heredity is carried within the cell nucleus, which is the control center not only in physical terms (it is usually located near the middle of the cell) but also because it contains the chromosomes. Within these threadlike structures is the genetic information organised in DNA molecules.

There are two basic types of cell in a multicellular organism: somatic, or body, cells, and germ, or reproductive, cells. The somatic cells are the primary components of most organisms, making up everything except some of the the cells in reproductive organs. The somatic cells of humans have 23 pairs of chromosomes, or 46 chromosomes overall, and are thus known as diploid cells. As the cells grow, they reproduce themselves by a process called mitosis, whereby a diploid cell splits to produce new diploids, each of which is a replica of the original. Thus cells grow and are replaced, making possible the formation of specific tissues and organs, such as muscles and nerves. Without mitosis, an organism's cells would not regenerate, resulting not only in cell death but possibly even the death of the entire organism. Mitosis is also the means of reproduction for organisms that reproduce asexually (see Reproduction).

A germ cell, by contrast, undergoes a process of cell division known as meiosis, whereby it becomes a haploid cell—a cell with half the basic number of chromosomes, which for a human would be 23 unpaired chromosomes. The sperm cells in a male and the egg cells in a female are both haploid germ cells: each contains only 23 chromosomes, and each is prepared to form a new diploid by fusion with another haploid. Sperm cells and egg cells are known as gametes, mature male or female germ cells that possess a haploid set of chromosomes and are prepared to form a new diploid by undergoing fusion with a haploid gamete of the opposite sex.

When egg and sperm fuse, they form a zygote, in which the diploid chromosome number is restored, with the zygote possessing the same chromosomes as both the sperm and the egg. This cell carries all the genetic information needed to grow into an embryo and eventually a full-grown human, with the specific traits and attributes passed on by the parents. Not all offspring of the same parents are the same, of course, and this is because the sperm cells and egg cells vary in their genetic codes—that is, in their DNA blueprints.

## The DNA Blueprint

To understand genes and their biological function in heredity, it is necessary to understand the chemical make up and structure of DNA. The complete DNA molecule often is referred to as the blueprint for life, because it carries all the instructions, in the form of genes, for the growth and functioning of organisms. This fundamental molecule is similar in appearance to a spiral staircase, which also is called a double helix. The sides of the DNA ladder are made up of alternate sugar and phosphate molecules, like links in a chain. The rungs, or steps, of DNA are made from a combination of four different chemical bases. Two of these, adenine and guanine, are known as purines, and the other two, cytosine and thymine, are pyrimidines. The four letters designating these bases—A, G, C, and T—are the alphabet of the genetic code, and each rung of the DNA molecule is made up of a combination of two of these letters.

## DNA Sequences

In this genetic code A always combines with T and C with G, to form what is called a base pair. Specific sequences of these base pairs make up the genes. Although a four-letter alphabet may seem rather small for constructing the extensive vocabulary that defines the myriad life-forms on Earth, in practice, the sequences of these base pairs make for almost limitless combinations. For any sequence, there are four possibilities as to the first two letters (AT, TA, CG, or GC) and four more possibilities for the second two letters. Thus, just for a four-letter sequence, there are 16 possibilities, and for each pair of letters added to the sequence, the total is multiplied by four. Given the long strings of base pairs that form DNA sequences, the numbers can be extremely large.

The more complex an organism, from bacteria to humans, the more rungs, or genetic sequences, appear on the ladder. The entire genetic make up of a human, for example, may contain three billion base pairs, with the average gene unit being 2,000-200,000 base pairs long. Each one of these combinations has a different meaning, providing the code not only for the type of organism but also for specific traits, such as brown hair and blue eyes, dimples, detached earlobes, and so on and on. Except for identical twins, no two humans have exactly the same genetic information.

## DNA Replication, Protein Synthesis, and RNA

Genetic information is duplicated during the process of DNA replication, which is initiated by proteins in the cells. To produce identical genetic information during cell mitosis, the DNA hydrogen bonds between the two strands arebroken, splitting the DNA in half lengthwise. This process begins a few hours before the initiation of cell mitosis, and once it is completed, each half of the DNA ladder is capable of forming a new DNA molecule with an identical genetic code. It can do this because of specific chemical catalysts (a substance that enables a chemical reaction without taking part in it) that help synthesise the complementary strand.

Catalysts formed from proteins are known as enzymes, and the functioning of specific cells and organisms is conducted by enzymes synthesised by the cells. Cells contain hundreds of different proteins, complex molecules that make up more than half of all solid body tissues and control most biological processes within and among these tissues. A cell functions in accordance with the particular protein—one of thousands of different types—it contains. It is the genetic base-pair sequence in DNA that determines, or "codes for," the specific arrangement of amino acids to build particular proteins.

Since the sites of protein production lie outside the cell nucleus, coded messages pass from the DNA in the nucleus to the cytoplasm, the material inside the cell that is external to the nucleus. This transfer of messages is achieved by RNA, or ribonucleic acid, and specifically by messenger RNA, or mRNA. Other types of RNA molecules are involved in linking the amino-acids together in a sequence form to shape the protein. (For more about amino acids, proteins, and enzymes, see the respective essays devoted to each subject.)

## Mutation

Once a protein has been created for a specific function, it cannot be changed. This is why the theory of acquired characteristics (the idea that changes in an organism's overall anatomy, as opposed to changes in its DNA, can be passed on to offspring) is a fallacy. People may have genes that make it easier for them to acquire certain traits, such as larger muscles or the ability to play the piano through exercise or practice, respectively, but the traits themselves, if they are acquired during the life of the individual and are not encoded in the DNA, are not heritable.

There is only one way in which changes that take place during the life of an organism can be passed on to its offspring, and that is if those changes are encoded in the organism's DNA. This is known as mutation. Suppose lung cancer develops in a man as a result of smoking; unless a tendency to cancer is already a part of his genetic make up, he cannot genetically pass the disease on to his unborn children. But if the tobacco has acted as a mutagen, a substance that brings about mutation, it is possible that his DNA can be altered in such a way as to pass on either the tendency toward lung cancer or some other characteristic.

Because DNA is extremely stable chemically, it rarely mutates, or experiences an alteration in its physical structure, during replication. But because there are so many strands of DNA in the world, and so much material in the strands, mutation is bound to happen eventually—and, to an extent, at least, this is a good thing. Mutation is the engine that drives evolution, and a certain amount of genetic variation is necessary if species are to adapt by natural selection to a changing environment. If it were not for mutation, neither humans nor the many millions of other species that exist would ever have appeared.

Mutation often occurs when chromosome segments from two parents physically exchange places with each other during the process of meiosis. This is known as genetic recombination. Genes also can change by mutations in the DNA molecule, which take place when a mutagen alters the chemical or physical make up of DNA. The mutations that result are of two types, corresponding to the two basic varieties of cell: somatic mutations, which occur solely within the affected individual, and germinal mutations, which happen in the DNA of germ cells, producing altered genes that may be passed on to the next generation.

The odd thing about mutations is that while most of them are harmful, the few that are beneficial are, as we have noted, the driving force behind the evolution of life-forms that successfully adapt to their environments. Thus, while most germinal mutations bring about congenital disorders (birth defects) ranging from physical abnormalities to deficiencies in body or mind to diseases, every once in a while a germinal mutation results in an improvement, such as a change in body coloring that acts as camouflage. If the trait improves an individual organism's chances for survival within a particular environment, it may become a permanent trait of the species, because the offspring with this gene have a greater chance of survival and thus will pass on the trait to succeeding generations. (For more about mutation, see the essay by that title. See also Evolution for a discussion of the role played by mutation and natural selection in the evolution of species.)

## The Genetics Revolution

In the modern world genetics plays a part in more dramatic breakthroughs than any other field of biological study. These breakthroughs have an impact in a wide variety of areas, from curing diseases to growing better vegetables to catching criminals. The field of genetics is in the midst of a revolution, and at the center of this exciting (and, to some minds, terrifying) phenomenon is the realm of genetic engineering: the alteration of genetic material by direct intervention in genetic processes. In agriculture, for instance, genes are transplanted from one organism to another to produce what are known as transgenic animals or plants. This approach has been used to reduce the amount of fat in cattle raised for meat or to increase proteins in the milk produced by dairy cattle. Fruits and vegetables also have been genetically engineered so that they do not bruise easily or have a longer shelf life.

Not all of the work in genetics is genetic engineering per se; in the realm of law, for instance, the most important application of genetics is genetic fingerprinting. A genetic fingerprint is a sample of a person's DNA that is detailed enough to distinguish it from the DNA of all others. The genetic fingerprint can be used to identify whether a man is the father of a particular child (i.e., to determine paternity), and it can be applied in the solving of crimes. If biological samples can be obtained from a crime scene—for example, skin under the fingernails of a murder victim, presumably the result of fighting against the assailant in the last few moments of life—it is possible to determine with a high degree of accuracy whether that sample came from a particular suspect. The use of DNA in forensic science is discussed near the conclusion of this essay.

## The Revolution in Medicine

Some of the biggest strides in genetic engineering and related fields are taking place, not surprisingly, in the realm of medicine. Genetic engineering in the area of health is aimed at understanding the causes of disease and developing treatments for them: for example, recombinant DNA (a DNA sequence from one species that is combined with the DNA of another species) is being used to develop antibiotics, hormones, and other disease-preventing agents. Vaccines also have been genetically re-engineered to trigger an immune response that will protect against specific diseases. One approach is to remove genetic material from a diseased organism, thus making the material weaker and initiating an immune response without causing the disease. (See Immunity and Immunology for more about how vaccines work.)

Gene therapy is another outgrowth of genetics. The idea behind gene therapy is to introduce specific genes into the body either to correct a genetic defect or to enhance the body's capabilities to fight off disease and repair itself. Since many inherited or genetic diseases are caused by the lack of an enzyme or protein, scientists hope one day to treat the unborn child by inserting genes to provide the missing enzyme. (For more about inherited disorders, see the essays Disease, Non-infectious Diseases, and Mutation.)

## The Human Genome Project

One of the most exciting developments in genetics is the initiation of the Human Genome Project, designed to provide a complete genetic map outlining the location and function of the 40,000 or so genes that are found in human cells. (A genome is all of the genetic material in the chromosomes of a particular organism.) With the completion of this map, genetic researchers will have easy access to specific genes, to study how the human body works and to develop therapies for diseases. Gene maps for other species of animals also are being developed.

The project had its origins in the 1990s, with the efforts of the United States Department of Energy (DOE) and the National Institutes of Health (NIH). The NIH connection is probably clear enough, but the DOE's involvement at first might seem strange. What, exactly, does genetics have to do with electricity, petroleum, and other concerns of the DOE? The answer is that the DOE grew out of agencies, among them the Atomic Energy Commission (AEC), established soon after the explosion of the two atomic bombs over Japan in 1945. Even at that early date, educated non-scientists understood that the radioactive fallout produced from nuclear weaponry can act as a mutagen; therefore, Congress instructed the AEC to undertake a broad study of genetics and mutation and the possible consequences of exposure to radiation and the chemical by-products of energy production.

Eventually, scientists in the AEC and, later, the DOE recognised that the best way to undertake such a study was to analyse the entire scope of the human genome. The project formally commenced on October 1, 1990, and is scheduled for completion in the middle of the first decade of the twenty-first century. Upon completion, the Human Genome Project will provide a vast store of knowledge and no doubt will lead to the curing of many diseases.

Still, there are many who question the Human Genome Project in particular, and genetic engineering in general, on ethical grounds, fearing that it could give scientists or governments too much power, unleash a Nazi-style eugenics (selective breeding) programme, or result in horrible errors, such as the creation of deadly new diseases. In fact, it is impossible to search "genetic engineering" on the World Wide Web without coming across the Web sites of literally dozens and dozens of agencies, activist groups, and individuals opposed to genetic engineering and the mapping of the human genome. For more about the Human Genome Project, genetic engineering, and their opponents, see Genetic Engineering.

## Genetics in Forensic Science

Forensic science, as we noted earlier, is the application of science to matters of law. It is based on the idea that a criminal always leaves behind some kind of material evidence that, through careful analysis, can be used to determine the identity of the perpetrator—and to exonerate someone falsely accused. Among those forms of material evidence of interest to forensic scientists working in the field of genetics are blood, semen, hair, saliva, and skin, all of which contain DNA that can be analysed. In addition, there are areas of forensic science that rely on biological study, though not in the area of genetics: blood typing as well as the analysis of fingerprints or bite marks, both of which have patterns that are as unique to a single individual as DNA is.

One of the first detectives to use science, including biology and medicine, in solving crimes was a fictional character: Sherlock Holmes, whose creator, the British writer Sir Arthur Conan Doyle (1859-1930), happened to be a physician as well.

The first full-fledged (and real) police practitioner of forensic science was the French police official Alphonse Bertillon (1853-1914), who developed an identification system that consisted of a photograph and 11 body measurements, including dimensions of the head, arms, legs, feet, hands, and so on, for each individual. Bertillon claimed that the likelihood of two people having the same measurements for all 11 traits was less than one in 250 million. In 1894 fingerprints, which were easier to use and more unique than body measurements, were added to the Bertillon system.

Fingerprints, unlike DNA, are unique to the individual; indeed, identical twins have the same DNA but different fingerprints. Mark Twain (1835-1910) could not have known this in 1894, when he published The Tragedy of Pudd'nhead Wilson, and the Comedy of Those Extraordinary Twins. Nonetheless, the story involves a murder committed by one man and blamed on his twin, who eventually is exonerated on the basis of fingerprint evidence—still a new concept at the time. In some situations, however, fingerprint evidence may be unavailable, and though law-enforcement agencies have developed extraordinary techniques for analysing nearly invisible (i.e., latent) prints, sometimes this is still not enough.

## The Simpson Case and the Controversy Over DNA Evidence

For example, in the infamous murder of Nicole Brown Simpson and Ron Goldman on June 12, 1994, fingerprint evidence would have been ineffective in the case against the suspect, the former football star and actor O. J. Simpson. Since Nicole Simpson was his ex-wife, the appearance of his prints at the scene of her murder in her Los Angeles home could be explained away easily, even though she had taken out a restraining order against her former husband (who she had accused of spousal abuse) some time before the murder. Rather than fingerprints, the prosecution in his murder trial used DNA evidence connecting blood at the crime scene with blood found in Simpson's vehicle. (Some of this blood was apparently his own, since he had mysterious cuts on his hands that he could not explain to police officers.)

A jury found Simpson not guilty on October 3, 1995, and jurors later claimed that the prosecution had failed to make a strong case using DNA evidence. Furthermore, they cited police contamination of the DNA evidence, which had been established in their minds by Simpson's defense team, as a cause for reasonable doubt concerning Simpson's guilt. In fact, assuming that the defense was fully justified in this claim, that would have meant only that the DNA samples would have been less (not more) likely to convict Simpson.

At the same time, a number of legitimate concerns regarding the use of DNA evidence were raised by experts for the defense in the Simpson trial. Samples can become contaminated and thus difficult to read; small samples are difficult for analysts to work with effectively; and results are often open to interpretation. Furthermore, the outcome of the Simpson case illustrates the fact that findings based on DNA evidence are not readily understood by non-specialists, and may not make the best basis for a case-particularly in one so fraught with controversy. The prosecution based its case almost entirely on extremely technical material, explained in excruciating detail by experts who had devoted their lives to studying areas that are far beyond the understanding of the average person. Attempting to wow the jurors with science, the prosecution instead seemed to create the impression that DNA evidence was some sort of hocus-pocus invented to frame an innocent man. Simpson went free, though the jury in a 1996 civil trial (which took a much simpler approach, eschewing complicated DNA testimony) found him guilty.

## DNA Evidence Success Stories

Because of the Simpson case, the use of DNA evidence gained something of a bad name. Nonetheless, it has been successful in less high profile cases, beginning in 1986, when English police tracked down a rapist and murderer by collecting blood samples from some 2,000 men. One of them, named Colin Pitchfork, paid another man to provide a sample in his place. This attracted the attention of the police, who tested his DNA and found their man.

Since that time, DNA evidence has been used in more than 24,000 cases and has aided in the conviction of about 700 suspects. The DNA in such cases is not always obtained from a human subject. In the investigation of the May 1992 murder of Denise Johnson in Arizona, a homicide detective found two seed pods from a paloverde tree in the bed of a pick up truck owned by the suspect, Mark Bogan. The accused man admitted having known the victim but denied ever having been near the site where her body was found. It so happened that there was a paloverde tree at the site, and testing showed that the DNA in the pods on his truck bed matched that of the tree itself. Bogan became the first suspect ever convicted by a plant.

On the other hand, in some cases, DNA evidence has cleared a suspect falsely accused. Such was the case with Kerry Kotler, convicted in 1981 for rape, robbery, and burglary and sentenced to 25-50 years in jail. In 1988, Kotler began petitioning for DNA analysis, which subsequently showed that his DNA did not match that of the rapist, who had left a semen sample in the victim's underwear. Kotler was released in December 1992 and in March 1996 was awarded $1.5 million in damages for his wrongful imprisonment. The story does not end there, however. Kotler's case turned out to be one of the more bizarre in the annals of forensic DNA testing. Perhaps he did not commit the first rape, but a month after he received the damage award, he was on his way back to prison for the August 1995 rape of another victim. This time prosecutors showed that Kotler's semen matched samples taken from his victim's clothing—and to prove their case, they used DNA testing.

# 2

# Fluorescent *In Situ* Hybridisation

## Introduction

FISH (Fluorescent *in situ* hybridisation) is a cytogenetic technique that can be used to detect and localise the presence or absence of specific DNA sequences on chromosomes. It uses fluorescent probes that bind to only those parts of the chromosome with which they show a high degree of sequence similarity. Fluorescence microscopy can be used to find out where the fluorescent probe bound to the chromosome. FISH is often used for finding specific features in DNA. These features can be used in genetic counselling, medicine, and species identification.

Probes are often derived from fragments of DNA that were isolated, purified, and amplified for use in the Human Genome Project. The size of the human genome is so large, compared to the length that could be sequenced directly, that it was necessary to divide the genome into fragments. The fragments were added into a framework that made it possible to use bacteria to replicate the fragments. The fragments were put into order by analysing size-exclusion separation of enzymatically-digested fragments. Clonal populations of bacteria, each population maintaining a single artificial chromosome, are stored in various laboratories around the world. The artificial chromosomes (BAC) can be grown, extracted, and labelled, in any lab. These fragments are on the order of 100 thousand base-pairs, and are the basis for most FISH probes.

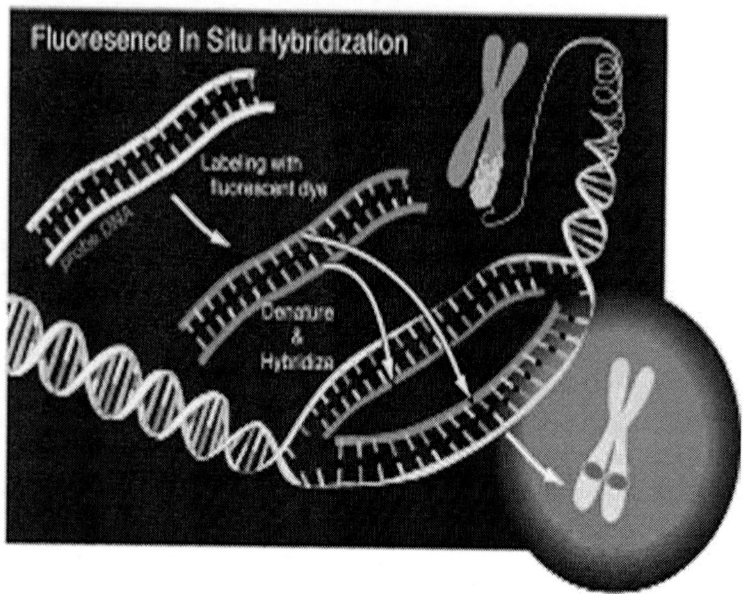


**Fig. 2.1:** Fluoresence *in Situ* Hybridisation

## Preparation and Hybridisation Process

First, a probe is constructed. The probe has to be long enough to hybridise specifically to its target (and not to similar sequences in the genome), but not too large to impede the hybridisation process, and it should be tagged directly with fluorophores, with targets for antibodies or with biotin. This can be done in various ways, for example nick translation and PCR using tagged nucleotides.

Then, an interphase or metaphase chromosome preparation is produced. The chromosomes are firmly attached to a substrate, usually glass. Repetitive DNA sequences must be blocked by adding short fragments of DNA to the sample. The probe is then applied to the chromosome DNA and incubated for ~12 hours while hybridising. Several wash steps remove all unhybridised or partially-hybridised probes. The results are then visualised and quantified using a microscope that is capable of exciting the dye and recording images.

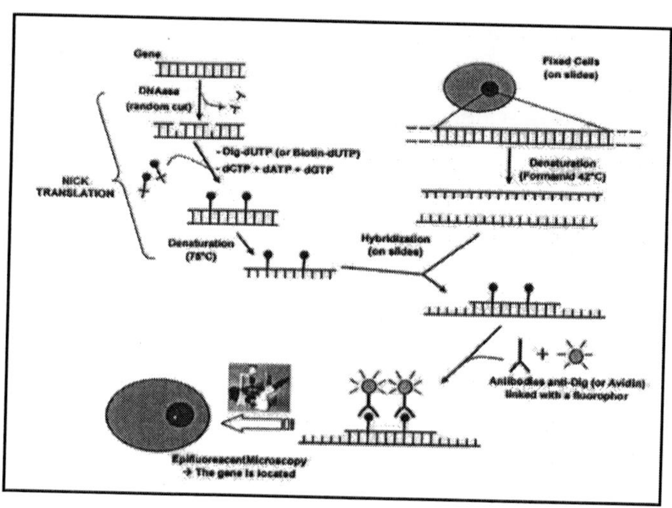


**Fig. 2.2:** Fish (Fluorescent *in Situ* Hybridisation

If the fluorescent signal is weak, amplification of the signal may be necessary in order to exceed the detection threshold of the microscope. The signal strength depends on many factors: Probe labelling efficiency, the type of probe, and the type of dye affect the fluorescent signal. Fluorescently-tagged antibodies or streptavidin are bound to the dye molecule. These secondary components are selected so that they have a strong signal.

## Fiber FISH

An alternative to interphase or metaphase preparations, fiber FISH, interphase chromosomes are attached to a slide in such a way that they are stretched out in a straight line, rather than being tightly coiled, as in conventional FISH, or adopting a random conformation, as in interphase FISH. This is accomplished by applying mechanical shear along the length of the slide, either to cells that have been fixed to the slide and then lysed, or to a solution of purified DNA. A technique known as chromosome combing is increasingly used for this purpose. The extended conformation of the chromosomes allows dramatically higher resolution—even down to a few kilobases. The preparation of fiber FISH samples, although conceptually simple, is a rather skilled art, and only specialised laboratories use the technique routinely.

## Variations on Probes and Analysis

FISH is a very general technique. It is often arbitrarily divided into more specific categories based on application, however, each category is similar in that, in a chemical sense, the technique is the same; hybridisation is the common denominator. The differences between the various FISH techniques are usually due to the construction and content of the fluorescently-labelled DNA probe. The size, overlap, colour, and mixture of the probes make possible all FISH techniques.

Probe size is important because longer probes hybridise more specifically than shorter probes. The overlap defines the resolution of detectable features. If the goal of an experiment is to detect the break point of a translocation, then the overlap of the probes — the degree to which one DNA sequence is contained in the adjacent probes — defines the minimum window in which the break point occurs.

The mixture of probes determines the type of feature the probe can detect. Probes that hybridise along an entire chromosome are used to count the number of a certain chromosome, show translocations, or identify extra-chromosomal fragments of chromatin. This is often called "whole-chromosome painting." If every possible probe is used, every chromosome, (in essence the whole genome) would be marked fluorescently, which would not be particularly useful for determining features of individual sequences. A mixture of smaller probes can be created that are specific to a particular region (locus) of DNA; these mixtures are used to detect deletion mutations. When combined with a specific colour, a locus-specific probe mixture is used to detect very specific translocations. Special locus-specific probe mixtures are often used to count chromosomes, by binding to the centromeric regions of chromosomes, which are unique enough to identify each chromosome.

Because modern microscopes can detect a range of colours in fluorescent dyes each human chromosome can be identified (M-FISH)using whole-chromosome probe mixtures and a variety of colours. There are currently twice as many chromosomes than fluorescent dye colours. However, ratios of probe mixtures can be used to create additional colours. As with comparative genomic hybridisation, the probe mixture for the secondary colours is created by mixing the correct ratio of two sets of differently-labelled probes for the same chromosome. Differently-coloured probes can be used for the detection of translocations. Several techniques exploit the resolution limitations of microscopes to resolve spatial distributions of dye below a few hundred nanometers. Colours that are adjacent appear to overlap, and a secondary colour is observed.

In reciprocal translocations, where both break points are known, locus-specific probes are made for it and part of the region one either side of break point. In normal cells, two colours will be visible; in diseased cells such as those found in BCR/ABL translocations, the two dye colours overlap, and a third colour is observed. This technique is known as double-fusion FISH or D-FISH. In translocations where only one of the break points is known or constant, locus-specific probes are made for one side of the break point and the other intact chromosome. In normal cells, the secondary colour is observed, but only the primary colour is observed when the translocation occurs. This technique is known as "break-apart FISH".

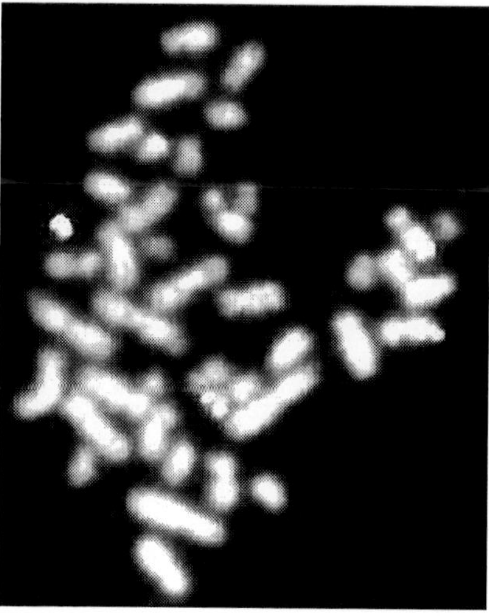

**Fig. 2.3:** A Metaphase Cell Positive for the BCR/ABL Rearrangement Using FISH

## Medical Applications

Often parents of children with a developmental delay want to know more about their child's conditions before choosing to have another child. These concerns can be addressed by analysis of the parents' and child's DNA. In cases where the child's developmental delay is not understood, the cause of it can be determined using FISH and cytogenetic techniques. Examples of diseases that are diagnosed using FISH include Prader-Willi syndrome, Angelman syndrome, 22q13 deletion syndrome, chronic myelogenous leukaemia, acute lymphoblastic leukaemia, Cri-du-chat, Velocardiofacial syndrome, and Down syndrome.

In medicine, FISH can be used to form a diagnosis, to evaluate prognosis, or to evaluate remission of a disease, such as cancer. Treatment can then be specifically tailored. A traditional exam involving metaphase chromosome analysis is often unable to identify features that distinguish one disease from another, due to subtle chromosomal features; FISH can elucidate these differences. FISH can also be used to detect diseased cells more easily than standard Cytogenetic methods, which require dividing cells and requires labor and time-intensive manual preparation and analysis of the slides by a technologist. FISH, on the other hand, does not require living cells and can be quantified automatically, a computer counts the fluorescent dots present. However, a trained technologist is required to distinguish subtle differences in banding patterns on bent and twisted metaphase chromosomes.

## Species Identification

FISH is often used in clinical studies. If a patient is infected with a suspected pathogen, bacteria, from the patient's tissues or fluids, are typically grown on agar to determine the identity of the pathogen. Many bacteria, however, even well-known species, do not grow well under laboratory conditions. FISH can be used to detect directly the presence of the suspect on small samples of patient's tissue.

FISH can also be to used compare the genomes of two biological species, to deduce evolutionary relationships. A similar hybridisation technique is called a zoo blot. Bacterial FISH probes are often primers for the 16s rRNA region.

FISH is widely used in the field of microbial ecology, to identify microorganisms. Biofilms, for example, are composed of complex (often) multi-species bacterial organisations. Preparing DNA probes for one species and performing FISH with this probe allows one to visualise the distribution of this specific species within the biofilm. Preparing probes (in two different colors) for two species allows to visualise/study co-localisation of these two species in the biofilm, and can be useful in determining the fine architecture of the biofilm.

## Lab-on-a-chip and FISH

Although interphase fluorescence in situ hybridisation (FISH) is a sensitive diagnostic tool used for the detection of chromosomal abnormalities on cell-by-cell basis, the cost-per-test and the technical complexity of current FISH protocols has inhibited its widespread utilisation. Lab-on-a-chip or microfluidic devices, incorporate networks of microchannels that can miniaturise, integrate and automate conventional analytical techniques onto chip-style platforms. Since microchannels permit sophisticated levels of fluid control (down to picolitres), these devices can reduce analysis times, lower reagent consumption, and minimise human intervention.

Currently, FISH has been performed on glass microfluidic platforms that standardise much of the protocol offering repeatable results that are accurate, cost-effective and easier to obtain in a clinical setting.

Compared to conventional FISH methods, these first implementations of on-chip FISH provide a 10-fold higher throughput and a 10-fold reduction in the cost of testing, enabling the simultaneous assessment of several chromosomal abnormalities or patients. It is increasingly essential that diagnostic tests determine the type and extent of chromosomal abnormalities for more informed diagnosis and for appropriate choice of treatment strategies. Since the on-chip FISH technique is 10 times more cost-effective than conventional methods, with the potential to be fully integrated and automated, this technology will make wide-spread genetic testing of patients more accessible in a clinical setting.

# 3

# Comparative Genomic Hybridisation (CGH)

## Introduction

Comparative Genome Hybridisation (CGH) is a technique used to visualise differences between two genomes. It holds promise for identifying genes that are amplified or deleted – useful for finding specific syndromes or examining differences throughout the entire genome. This article discusses the development, mechanism, and application of CGH.

Chromosomal abnormalities are one cause of phenotypic differences between normal and affected individuals. The technique of chromosomal banding was developed in the early 1970s to identify these abnormalities. Banding uses chemicals that alter the colorimetry of DNA pairs to produce differently stained regions on chromosomes. These regions appear as bands of varying shades of grey under light or fluorescence microscopy. This allows the detection of structural rearrangements such as translocations, deletions, duplications, polymorphic variations, or ploidy differences. However, with time it has become evident that routine banding methods cannot uncover all abnormalities, as only rearrangements greater than 10 megabases can be reliably detected. As a result, numerous technologies have been developed to provide a higher resolution in the delineation of subtle rearrangements.

Molecular cytogenetics was developed in the 1980s and is based on the hybridisation of DNA probes to target DNA. These probes are labelled using fluorochromes for a variety of different regions of the chromosome, including the centromere, telomere, specific genes, or even the entire chromosome. Application of fluorescence *in situ* hybridisation (FISH) in conjunction with routine banding vastly improved the detection of changes in DNA. FISH is a popular molecular cytogenetic technique that hybridises the labelled probe to the target DNA, through a series of denaturing and re-annealing steps. Currently there are commercially available probe kits for common aneuploidies, specific deletions, duplications, fusion gene rearrangements in cancers, and telomeric regions. Though this technology provides a higher resolution of 5 to 10 megabases, the largest limitation is that one must decide beforehand on specific regions in which to look.

## Technology of Genome Hybridisation

A new technology called Comparative Genome Hybridisation (CGH) has a resolution power equivalent to FISH. Although CGH is limited to using metaphase chromosomes, it has many other advantages. CGH compares differences in two different genomes; DNA is extracted from both the test subject and a normal reference subject. The two samples are then labelled, usually green (cyanine 3, or Cy3) and red (cyanine 5, or Cy5). The ratio of the two fluorochromes on the metaphase chromosomes are then compared. To illustrate its application, consider a red test sample and a green reference sample. If a region is amplified in the test sample, the corresponding region of hybridised chromosome appears red. However, if a region is deleted in the test sample, the corresponding region of the hybridised chromosome would appear green. The ratio of the test to reference fluorochromes is quantified by a computer mapped to each chromosome pair and visualised with a digital image.

The major advantage of this technology is that it does not require a prior hypothesis of a genetic defect, and can thus identify regions of genetic imbalance. CGH is especially useful in detecting deletions, duplications, non-reciprocal translocations, and gene amplifications throughout the entire genome. However, CGH does come with limitations such as failure to detect balanced translocations and inversions due to incomplete genomic information. Also, CGH analysis does not provide possible sites of imbalance, thus locating the site of the imbalance can be a concern without the availability of routine cytogenetics methods.

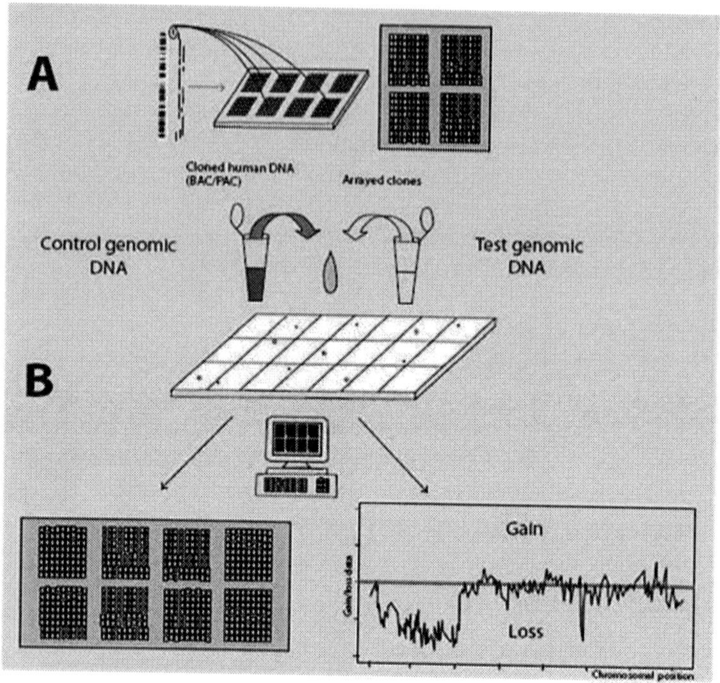


Fig. 3.1: A. Clones of all the genes on a chromosome or entire genome are printed onto a glass microscope slide (arrayed). The right side shows how they can be stained to show the morphology and placement of each "spot" of the cloned DNA

B. The reference (left) and test (right) are labelled with different fluorochromes, and then mixed on the array. Computer imaging shows a yellow hybridisation colour for the clones that are equal in proportion, green for those genes deficient in the test, and red for those in excess in the test DNA (lower left). A plot of ratio between the reference and test DNA for each clone showing dosage differences (lower right).

## Routine Cytogenetics Methods

CGH has been improved through use with microarrays. For Array CGH (aCGH), large numbers of mapped clones (cDNA,

Bacterial Artificial Chromosomes, PCR-generated sequences, or oligonucleotides) spotted onto a glass slide are used instead of metaphase chromosomes. This increases the resolution of screening for gains or losses of genomic copy numbers so that the resolution is only limited by the size and density of the target sequence. Typically, one clone is used per megabase. The test and reference genomes are labeled and co-hybridised onto a microarray. The array is imaged and the fluorescence intensities are calculated for each mapped clone, with the resulting ratio reflecting the differences in DNA copies. Computer imaging reveals a yellow hybridisation colour for all clones that are in equal proportion between the test and reference. Clones that are deleted in the test DNA will appear green, and those duplicated will appear red. A computerised plot of the ratio between test and reference is made to reveal dosage differences. It is visualised as a deviation of the ratio from zero.

Array CGH is advantageous because of its higher resolution and dynamic range, and there is direct mapping of aberrations to the genome sequence because the alterations are immediately linked to genomic markers via the clones used. aCGH is also advantageous because it can be automated for high-throughput application, and it has a low false-positive count. There is also wide flexibility with the type of array available, ranging from whole-genome arrays to specialised arrays for specific diseases (Spiecher and Carter, 2005). The highest resolution aCGH is provided by oligonucleotide arrays, produced by spotting oligonucleotides onto a slide, or synthesising them directly onto the glass. This allows for a resolution as specific as 15 to 20 kilobases. It is even possible to find single nucleotide polymorphisms (SNP) using SNP arrays, which are high-density oligonucleotide-based arrays. These are useful in identifying Loss of Heterozygosity which is the loss of a single allele at a given locus as a result of genomic mutations (Spiecher and Carter, 2005). The primary drawback of aCGH is the cost; routine testing of the equipment has been quoted at $90 000, and each array costs more than $1 200 in some cases (Kolomietz, 2006).

Currently CGH has been used for the analysis of gains or losses in tumours, but remains largely a tool for research purposes. CGH has also been important in identifying the presence and levels of normal genomic variation, which are assumed to be responsible for individual differences in gene expression, phenotypic variation, and susceptibility to disease (Spiecher and Carter, 2005).

## Application of Genetic Techniques

There are no genetic techniques that can be used to find every variation. For example, routine G-Banding has been used as the standard for decades in screening for polyploidy, aneuploidy, or rearrangements. However, its low resolution makes it challenging to detect microdeletions, microduplications, and subtle rearrangements; it is also reliant on the adequate preparation and morphology of the chromosomes. FISH can be quite effective for clinical laboratories in identifying specific syndromes, provided that FISH probes are commercially available (Spiecher and Carter, 2005). CGH is very useful for searching the entire genome and for indicating the presence of genetic abnormalities, but remains very expensive, allowing only well-funded labs to utilise CGH technology.

Each technique described seems to be useful under different circumstances, while lacking in terms of resolution or economics. At this time, it is best to combine techniques in order to provide a comprehensive analysis, if the tests are available. An example given by Speicher and Carter (2005) illustrates this point best using the example of a child exhibiting mental disability and dysmorphic features due to an unknown chromosomal rearrangement. First, G-banding was able to show a complex chromosomal rearrangement, but could not identify the structural chromosomal rearrangements. FISH could identify the rearrangement involving the chromosomes 2, 5, 6, 8, and 14, but could not tell anything about the imbalance. Lastly, array CGH was able to identify four genomic deletions, and allowed direct mapping of the deletion breakpoints onto the reference. The number of genes involved could be determined using internet based genome browsers (Spiecher and Carter, 2005). With strategies similar to this, it is plausible that the origin of some genetic disorders can be confirmed. Also, the possibility of developing a more cost efficient technology will allow genetic disorders to be diagnosed in the most economic manner under the tight restrain of health care dollars.

# 4

# Molecular Cytogenetics Techniques

## Inroduction

As the pioneer among molecular cytogenetics techniques, fluorescence in situ hybridisation (FISH) allows identification of specific sequences in a structurally preserved cell, in metaphase or interphase. This technique, based on the complementary double-stranded nature of DNA, hybridises labelled specific DNA (probe). The probe, bound to the target, will be developed into a fluorescent signal. The fact that the signal can be detected clearly, even when fixed in interphase, improves the accuracy of the results, since in some cases it is extremely difficult to obtain mitotic samples. FISH is still used mostly in research, but there are diagnostic applications. New nomenclature is being developed in order to define many of the aberrations that were not distinguished before FISH. Prenatal diagnosis of aneuploidies and malignancies are promptly detected with FISH, which is very useful in critical cases. In some tumors, where chromosomal abnormalities are too complicated to classify manually, the technique of comparative genomic hybridisation (CGH), a competitive FISH, allows examiners to determine complete or partial gain or loss of chromosomes. CGH results allow the classification of many tumor cell lines and along with other complementary techniques, like microdissection-FISH, PRINS, etc., increase the possibility of choosing an appropriate treatment for cancer patients.

When the improvements in cytogenetics techniques seemed to have reached a plateau, a combination of cytogenetics and molecular biology gave rise to fluorescence *in situ* hybridization (FISH) opening many new opportunities in the field.

The main reasons for this were: (1) It opens the possibility of obtaining cytogenetic results from interphase cells, impossible to achieve prior to the FISH technique; (2) Its colourful results are easy to interpret, explain and understand; (3) DNA is a very stable molecule, therefore archived material is suitable to be analysed and paraffin-embedded tissues can be studied; (4) It can be combined with other cytogenetics and molecular techniques, complementing and improving the outcome of several other methods.

Molecular cytogenetics are extensively used in many fields. Among these are: radiation biology, somatic cell genetics (cancer), prenatal and postnatal genetics, and preimplantational diagnosis. Some other fields are less explored, such as microbiology, ecology and evolution. Molecular cytogenetics techniques will have many new applications in probe synthesis for animal research models, and for specific diagnosis in economically important plant species, farm animals and pets.

## Applications of Molecular Cytogenetics

### *Radiation Biology*

FISH is one of the latest technologies used in radiation biology and it has brought many new insights to old problems such as basic research on aberration formation mechanisms. Initial chromosome damage and repair systems involve radiobiodosimetry.

Radiation-induced aberrations are one of the parameters analysed after nuclear accidents. However, it is possible to estimate the risk the victims will have to confront and take preventive measures accordingly. Radiation-induced aberrations can be symmetrical or asymmetrical. The asymmetrical type aberrations (those which have an acentric fragment: dicentric, interstitial deletion, asymmetrical exchange, etc.) are lethal for the cells. Acentric fragments cannot be attached to the spindle in mitosis and usually they disappear, so cells carrying them will lack genetic information and are eliminated from the cell population a few generations post-irradiation. On the other hand, symmetrical type aberrations also decrease with post-irradiation time, but since they do not carry an acentric fragment they can survive as a clone in bone marrow, and therefore we can find them many years after exposure.

In whole body irradiation, we can estimate the exposure with a calculation of aberration frequency. FISH can be used for some of the chromosomes and then extrapolated to give the whole picture.

Radiation is also used for medical purposes, and it is not uncommon to find patients who overreact to routine doses. On the other hand, some tumours are highly resistant to radiation therapy. Chromosome aberration frequencies, studied *in vivo* after irradiation of tumours, have been reported as a tool to distinguish tumor sensitivity or resistance to radiation. A non-invasive technique, that consists of aberration scoring in peripheral blood, is being developed in order to detect possible overreactions in patients that need radiotherapy. The same system is used to detect cancer-prone patients (Figure 4.1). For these studies, aberration scoring with plain Giemsa staining does not seem to have enough sensitivity. There have been some efforts to improve this method with FISH painting. This methodology has already been used in tumour cells).

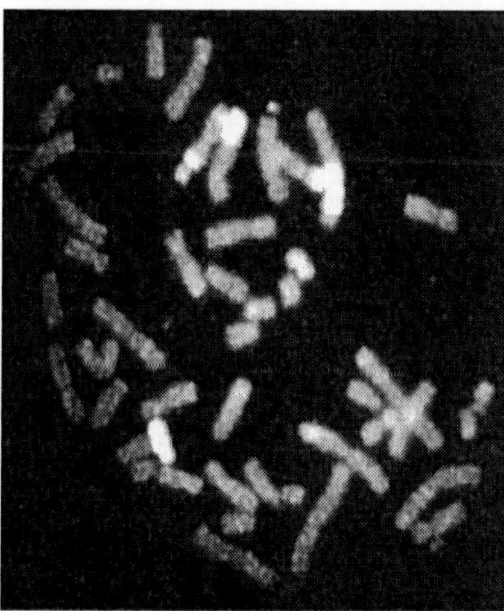

**Fig. 4.1:** Irradiated human lymphocyte. X chromosome "painted" yellow shows insertions and translocations induced *in vitro* by 7-Gy gamma rays.

## Somatic Cell Genetics

In somatic cell genetics we can study features related to chromosome topology, nuclear structure and gene mapping , as well as different malignancies and abnormalities related to cancer in hematological alterations and solid tumours.

Chromosomal aberrations are found in cancer and tumour cell lines. Some of them are already characterised and correlated with specific syndromes (*See Table 4.1 on page 33 and Table 4.2 on page 34)*) and some of them have yet to be associated with a clinical outcome. The continuous input of data and reports connecting clinical and biochemical data with chromosomal and genetic data will increase the list of known genes involved in each malignancy. The great advantage of the molecular cytogenetics techniques in these cases is the possibility to study interphase cells (*See Fig. 4.2A on page 35*). A good example is the amplification of the Her/2 neu oncogene (*See Fig. 4.2B on page 35*). About 30 per cent of breast tumours (and other tumours) have amplification of this oncogene. These tumours have a bad prognosis, because they usually do not respond to the normal chemotherapeutic protocol and they are very aggressive. This oncogene codes for a membrane growth factor receptor. A humanised antibody against the specific receptor has recently been approved by the FDA and improves the condition of these patients considerably. Now they have an alternative. This is a typical example in which we can count on interphase cytogenetics of archived paraffin-embedded tissues.

Chromosome triploidy involving chromosome 12 is a common feature of some leukaemias. The fact that FISH with centromere probes allows us to count enough cells to make a statistical analysis of interphase cells makes the study of minimal residual disease (MRD) much faster and less demanding than by classical karyotyping.

## Advanced Molecular Cytogenetics in Neoplasias

Interphase cytogenetics is indeed very useful when we know what we are looking for. But there is another problem in genetically unstable solid tumours. After the initial mutation some tumours become unstable and soon they have a wide range of different chromosomal aberrations. In any of these cases, comparative genomic hybridisation (CGH) is a very useful tool.

**Table 4.1**

| Pathology | Detectable anomaly | Application |
|---|---|---|
| Chronic myelogenous leukaemia<br>Acute myeloid leukaemia | Translocation 9, 22<br>BCR/ABL genes<br>Trisomy 8<br>Trisomy 19<br>Trisomy 21<br>Iso (17q) | Diagnosis<br><br>Prognosis |
| Chronic myelogenous leukaemia<br>Acute lymphocytic leukaemia<br>Myeloproliferative disorders<br>Myelodysplastic syndromes<br>Chronic lymphocytic leukaemia<br>Polycythemia vera | Trisomy 8<br>8(p11.1-q11.1) | Prognosis<br>Myeloid blast crisis and basophilia |
| | | |
| Acute lymphocytic leukaemia<br>Therapy-related chronicity<br>Myelogenous leukaemia<br>Myelodysplastic syndromes | Deletion 5q31<br>ERG-1 | Prognosis |
| Acute promyelogenous leukaemia | Translocation 15, 17<br>PML/RARA<br>3'RAR/5'RARA | Diagnosis<br>Prognosis<br>Early chemotherapy in APL is critical |
| Myelodysplastic syndromes | 7q31 D 75486 | Follow-up |
| B-cell chronic lymphocytosis leukaemia | Deletion 13q<br>D13525 13q14.3q<br>P53 Deletion 17p13.1 | Prognosis<br>Therapy resistance |
| Childhood B-cell acute lymphocytic leukaemia | Translocation 12, 21<br>TEL/AML1 | Prognosis<br>Checking for MRD |
| Chronic lymphocytic leukaemia<br>Acute lymphoblastic leukaemia<br>Follow-up in bone marrow transplant<br>Minimal residual disease (MRD) | Trisomy 12p11-q11<br>Isochromosome 17<br>17q11 q12<br><br>Probes for X & Y | Prognosis<br>Myeloid blast crisis and basophilia |

**Table 4.2**

| *Pathology* | *Detectable anomaly* | *Application* |
|---|---|---|
| Retinoblastoma | RBI gene: 13q14<br>D13525 | Diagnosis |
| Breast cancer | Deletion 7q31-D75522<br>Amplification 17q11.2q12<br>Her-2/neu<br>Amplification 8q24.2-q24.3<br>Oncogene c-Myc<br>Amplification 20q13.2<br>Amplification 11q13 | Prognosis<br>Therapy resistance |
| Neuroblastoma | Amplification N-myc<br>2p23-p24 | Prognosis<br>Evolution |
| Carcinoma of the colon<br>Osteosarcoma | Deletion 13q14<br>RB-1 | Diagnosis-prognosis<br>Treatment indicator |
| Recurrent prostate tumours | Amplification Xq12<br>Androgen receptor locus | Diagnosis-prognosis<br>Treatment indicator |

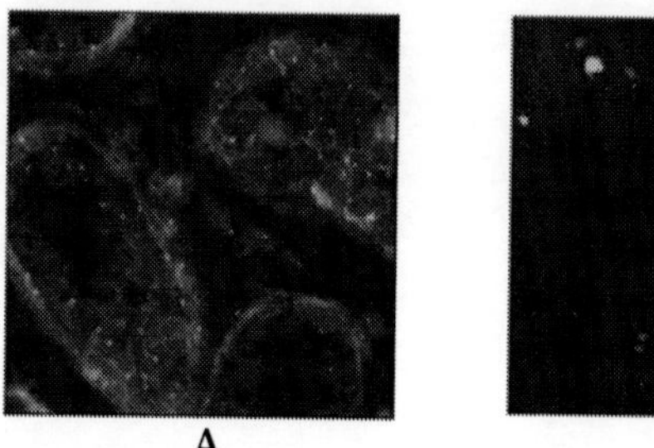


Fig. 4.2: Molecular cytogenetics techniques in interphase cells (A) and Her/2 neu oncogene (B). A. Paraffin-embedded tissue (60X): disgenesic gonads, X centromere (green), Y centromere (red). B. Paraffin-embedded tissue (100X): Her/2 neu oncogene (red) amplified in breast cancer control hybridisation (centromere 17 green).

CGH is a special way to perform FISH (Figure 4.3A, B). It is a competition between normal cell DNA and suspected tumour DNA that allows the detection of deletions or amplifications that correspond to the entire tumour cell population. As with any methodology CGH complements other studies because by itself it will not detect balanced translocations. Nevertheless it is a very useful tool, because it allows us to recognise unbalanced aberrations, which are a common feature of the whole tumour.

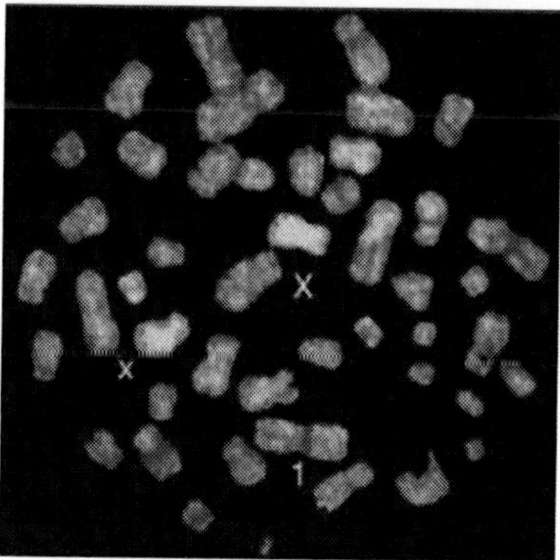


Fig. 4.3A: Human metaphase in corporative genomic hybridisation (CGH), result of a multiple X case. X and x chromosomes with a brighter hybridisation signal.

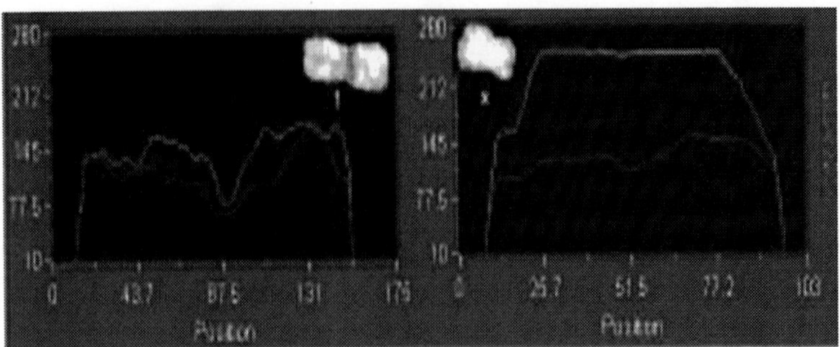

**Fig. 4.3B:** Comparative genomic hybridisation analysis examples. Left, Chromosome one: normal signal with green/red ratio about one. Right, Chromosome X with green/red ratio about, 2.5, meaning supernumerary.

Homogenously stained regions (HSR) and double minutes are forms of DNA amplification that appear in some types of malignancies. The microdissection of chromosomes (Figure 4.4) and subsequent probe development allow us to map a specific HSR to the DNA of origin. By making a probe from these microdissected areas we can use a normal cell as a target and find out where the amplified DNA comes from.

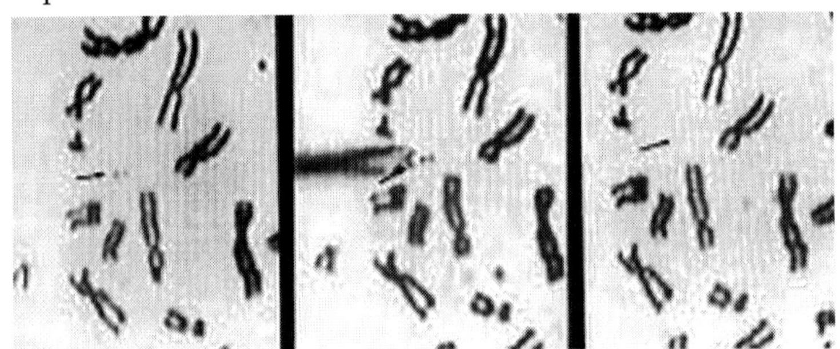

**Fig. 4.4:** Microdissection procedure. Left, Selection of a chromosome piece for microdissection. Middle, Microneedle approaching the selected piece. Right, Area post-microdissection.

## Hereditary Genetics

Hereditary genetics includes syndromes passed on from the progenitors or that develop spontaneously during conception or embryogenesis.

*Prenatal testing.* The fact that FISH allows us to perform interphase cytogenetics makes it suitable for very rapid prenatal diagnosis. Specific probes for the most common alterations can be used, giving a fast result while the regular cytogenetic results are still being processed. Even more important is that the probability of detecting any mosaicism for the probes that one is using is incremented, since metaphases and interphases are included in the scoring.

*Postnatal, microdeletion syndromes.* As mentioned before, there is no one methodology that can resolve all cases. Some syndromes and specific mutated loci can be detected or diagnosed by PCR, but in cases where there is a microdeletion involved, PCR is not useful since a quantitative PCR is very complicated. Specific locus FISH probes can give an answer. In these cases we have one chromosome in which a single locus is deleted, and the other is normal (Figure 4.5). Classical cytogenetics cannot detect these deletions, because they are microdeletions with only a few hundred base pairs (*See Table 4.3 on next page*).

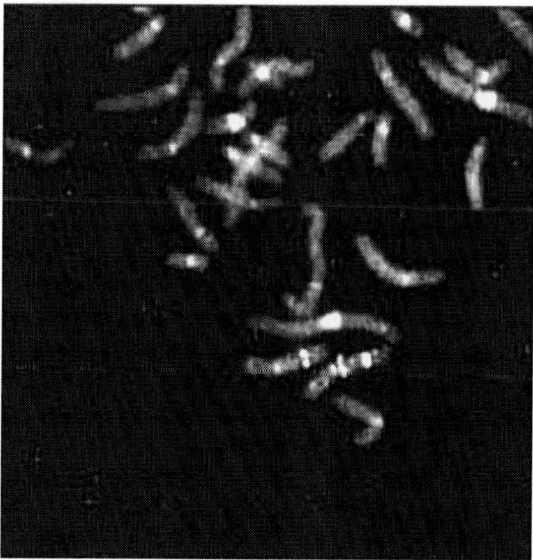

**Fig. 4.5:** Williams syndrome human metaphase. Detection of chromosome 7 microdeletion: elastin gene (red). Green chromosome 7 centromere is a control for hybridisation.

**Table 4.3**

| *Syndrome* | *Locus* |
|---|---|
| Prader-Wi)li/Angelman | 15q11-q13//15p11.2//15q22SNRPNyD15S10 |
| Velocardiofacial (Catch 22) | 22q11-22q13.3 |
| Williams syndrome (Elastia gene) | 7q11.23/7q31 |
| Kallman's syndrome | Xp22.3KAL |
| Cri-du-chat | D5S23 5p15.2 |
| Steroid sulfatase | Xp22.3STS |
| Smith-Magenis | 17q11.2 |
| Miller Diecker | 17p12.3 |
| Wolf-Hirshborn | 4p16.3 |

*Preimplantational genetics.* Some phenotypically normal persons carry balanced translocations or other types of aberrations. They usually have trouble having children or if they do, frequently their children have serious health problems. In other cases the abnormalities are spontaneous cases, but the parents are afraid to repeat the bad experience and seek prenatal testing for new pregnancies. In any case, waiting for the results causes considerable anxiety that can be minimised by using molecular cytogenetics techniques, such as FISH or PRINS.

As FISH can be performed with a single cell, preimplantation diagnosis is possible (Figure 4.6). By knowing a priori what to look for, a combination of centromere probes can be used to detect aneuploidies . When the female carries the translocation, pre-ICSI determination of the quality of the oocyte can be done with the polar body . If the male carries the translocation, studies on the disjunction of the sperm can determine the probability that a single sperm is normal.

Other less developed applications (Figure 4.7) but with just as promising a future are:

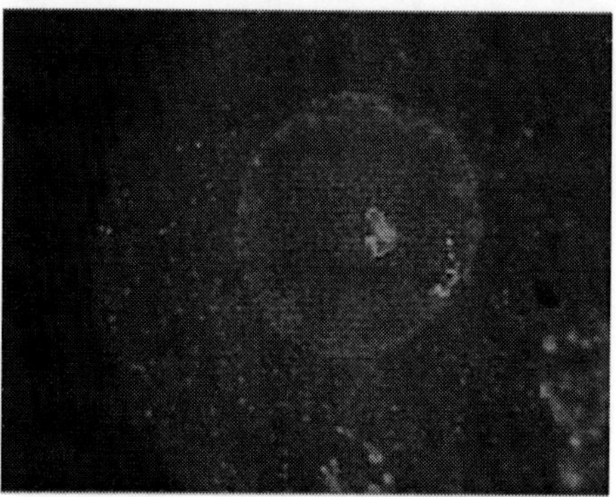

**Fig. 4.6:** Human blastomere hybridised with total chromosome painting of chromosome painting of chromosome 21: two chromosome 21 domains in red

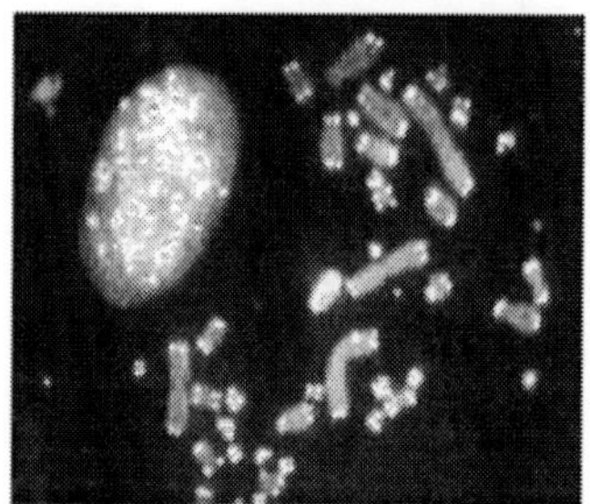

Human telomeres hybridized in a turtle Trachemys scripta metaphase.

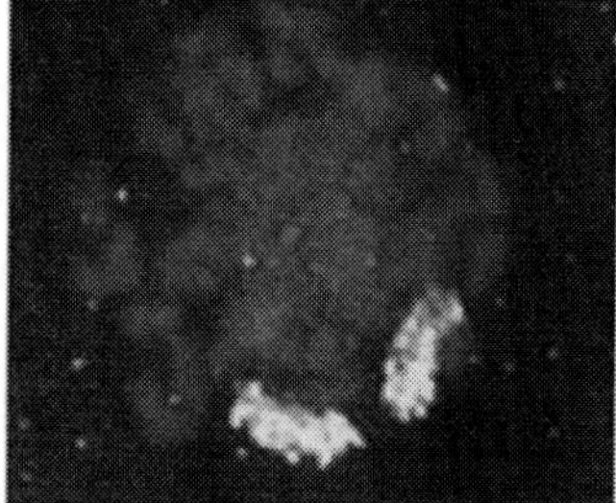

Pair one of Pati-fish chromosomes (kindly provided by Alejandro Laudicina).

**Fig. 4.7:** Applications of FISH probes in different species

## Microbiology

This area is perhaps the least developed but it further illustrates the multiple uses of molecular cytogenetics . Diagnoses of the presence of certain viruses has been reported using probes for specific DNA of the infectious agent.

## Ecology

Aberrations are an end point for genetic damage. Radiation by itself, or potentiated by other contaminants, can cause slight genetic damage. This might not affect phenotypic expression, but for example, translocation will decrease the pregnancy rate, and this in turn will slowly decrease the population growth of some species in the contaminated area. People might also be living in the surrounding areas, and be exposed to the same damage, affecting their life in the long run or even the next generation.

## Evolution

During the speciation process in evolution some of the chromosome structures are conserved and some of them are interchanged with each other. The FISH technique has proved that related species not only have common genes but also common fragments of chromosomes .

## Agriculture and Animal Husbandry

Chromosomal aberrations were first analysed by banding techniques alone. Classical cytogenetics is based on species-specific bands to interpret the chromosomal exchanges that lead to aberrations. Training for these banding techniques is quite difficult and, every species has its own banding pattern, which the technician needs to learn. This is even more difficult for the many species for which banding does not work well and there is little background information in the literature to help with the classification of chromosomes.

Once an appropriately labelled DNA probe is available, FISH becomes the technique of choice (Figure 4.8). That is why FISH has been so useful in different fields.

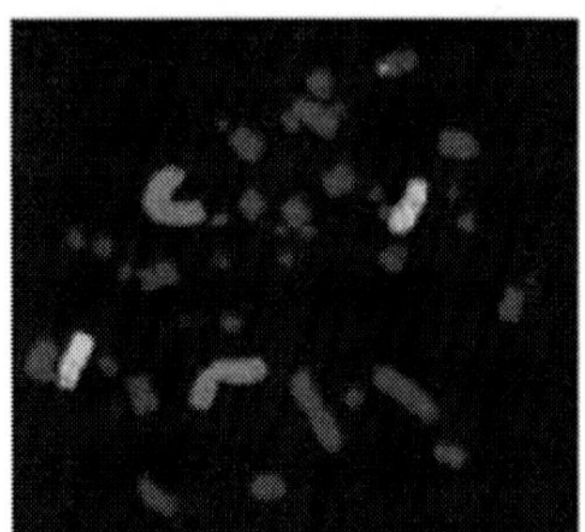

**Pair one (red) and three (green) of turtle Trachemys scripta metaphase**

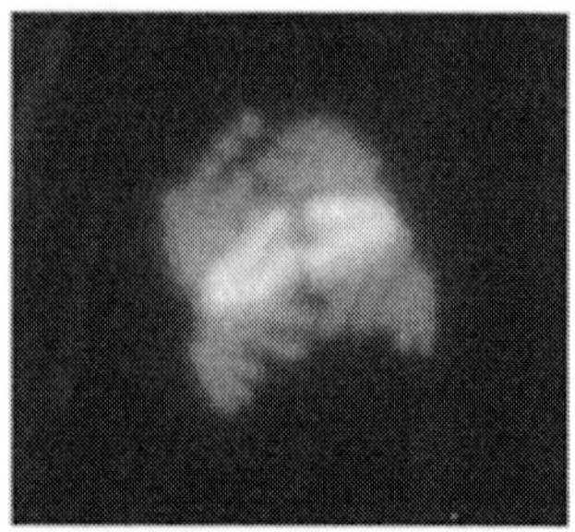

**Plasmid (yellow) stably introduced in an Aedes albopictus mosquito cell (red) as a new chromosome**

**Fig. 4.8**

Molecular cytogenetics has gone a long way in a short time. Efficiency and sensitivity have been improved by combining methodologies. The FDA has already approved some of diagnostic testing techniques and more are under study for their approval. Still, many probes need to be developed, for more human specific genes and chromosome fragments, as well as for non-human species. Our group has been developing projects in this direction.

# 5

# General Cytogenetics Information

## Introduction

Cytogenetics is the study of chromosomes and the related disease states caused by abnormal chromosome number and/or structure. Chromosomes are complex structures located in the cell nucleus, they are composed of DNA, histone and non-histone proteins, RNA, and polysaccharides. They are basically the "packages" that contain the DNA. Normally chromosomes can't be seen with a light microscope but during cell division they become condensed enough to be easily analysed at 1000X. To collect cells with their chromosomes in this condensed state they are exposed to a mitotic inhibitor which blocks formation of the spindle and arrests cell division at the metaphase stage.

A variety of tissue types can be used to obtain chromosome preparations. Some examples include peripheral blood, bone marrow, amniotic fluid, and products of conception. Although specific techniques differ according to the type of tissue used, the basic method for obtaining chromosome preparations is as follows:

- Sample log-in and initial setup.
- Tissue culture (feeding and maintaining cell cultures).
- Addition of a mitotic inhibitor to arrest cells at metaphase.
- Harvest cells. This step is very important in obtaining high quality preparations. It involves exposing the cells to a hypotonic solution followed by a series of fixative solutions.

This causes the cells to expand so the chromosomes will spread out and can be individually examined.

- Stain chromosome preparations to detect possible numerical and structural changes.

## Chromosome Morphology

Under the microscope chromosomes appear as thin, thread-like structures. They all have a short arm and long arm separated by a primary constriction called the centromere. The short arm is designated as *p* and the long arm as *q*. The centromere is the location of spindle attachment and is an integral part of the chromosome. It is essential for the normal movement and segregation of chromosomes during cell division. Human metaphase chromosomes come in three basic shapes and can be categorised according to the length of the short and long arms and also the centromere location. Metacentric chromosomes have short and long arms of roughly equal length with the centromere in the middle. Submetacentric chromosomes have short and long arms of unequal length with the centromere more towards one end. Acrocentric chromosomes have a centromere very near to one end and have very small short arms. They frequently have secondary constrictions on the short arms that connect very small pieces of DNA, called stalks and satellites, to the centromere. The stalks contain genes which code for ribosomal RNA.

## Chromosome Analysis

Virtually all routine clinical Cytogenetic analyses are done on chromosome preparations that have been treated and stained to produce a banding pattern specific to each chromosome. This allows for the detection of subtle changes in chromosome structure. The most common staining treatment is called G-banding. A variety of other staining techniques are available to help identify specific abnormalities. Once stained metaphase chromosome preparations have been obtained they can be examined under the microscope. Typically 15-20 cells are scanned and counted with at least 5 cells being fully analysed. During a full analysis each chromosome is critically compared band-for-band with it's homolog. It is necessary to examine this many cells in order to detect clinically significant mosaicism.

Following microscopic analysis, either photographic or computerised digital images of the best quality metaphase cells are made. Each chromosome can then be arranged in pairs according to size and banding pattern into a karyotype. The karyotype allows the Cytogeneticist to even more closely examine each chromosome for structural changes. A written description of the karyotype which defines the chromosome analysis is then made.

## Normal Chromosomes

Normal human somatic cells have 46 chromosomes: 22 pairs, or homologs, of autosomes (chromosomes 1-22) and two sex chromosomes. This is called the diploid number. Females carry two X chromosomes (46,XX) while males have an X and a Y (46,XY). Germ cells (egg and sperm) have 23 chromosomes: one copy of each autosome plus a single sex chromosome. This is referred to as the haploid number. One chromosome from each autosomal pair plus one sex chromosome is inherited from each parent. Mothers can contribute only an X chromosome to their children while fathers can contribute either an X or a Y.

- Normal 550 and 650 Band Karyotypes

## Chromosome Abnormalities

Although chromosome abnormalities can be very complex there are two basic types: numerical and structural. Both types can occur simultaneously.

Numerical abnormalities involve the loss and/or gain of a whole chromosome or chromosomes and can include both autosomes and sex chromosomes. Generally chromosome loss has a greater effect on an individual than does chromosome gain although these can also have severe consequences. Cells which have lost a chromosome are monosomy for that chromosome while those with an extra chromosome show trisomy for the chromosome involved. Nearly all autosomal monosomies die shortly after conception and only a few trisomy conditions survive to full term. The most common autosomal numerical abnormality is Down Syndrome or trisomy-21. Trisomies for chromosomes 13 and 18 may also survive to birth but are more severely affected than individuals with Down Syndrome.

Curiously, a condition called triploidy in which there is an extra copy of every chromosome (69 total), can occasionally survive to birth but usually die in the newborn period.

Another general rule is that loss or gain of an autosome has more severe consequences than loss or gain of a sex chromosome. The most common sex chromosome abnormality is monosomy of the X chromosome (45, X) or Turner Syndrome. Another fairly common example is Klinefelter Syndrome (47, XXY). Although there is substantial variation within each syndrome, affected individuals often lead fairly normal lives.

Occasionally an individual carries an extra chromosome which can't be identified by it's banding pattern, these are called marker chromosomes. The introduction of FISH techniques has been a valuable tool in the identification of marker chromosomes.

Structural abnormalities involve changes in the structure of one or more chromosomes. They can be incredibly complex but for the purposes of this discussion we will focus on the three of the more common types:

- Deletions involve loss of material from a single chromosome. The effects are typically severe since there is a loss of genetic material.
- Inversions occur when there are two breaks within a single chromosome and the broken segment flips 180° (inverts) and reattaches to form a chromosome that is structurally out-of-sequence. There is usually no risk for problems to an individual if the inversion is of familial origin (has been inherited from a parent.) There is a slightly increased risk if it is a de novo (new) mutation due possibly to an interruption of a key gene sequence. Although an inversion carrier may be completely normal, they are at a slightly increased risk for producing a chromosomally unbalanced embryo. This is because an inverted chromosome has difficulty pairing with it's normal homolog during meiosis, which can result in gametes containing unbalanced derivative chromosomes if an unequal cross-over event occurs.

- Translocations involve exchange of material between two or more chromosomes. If a translocation is reciprocal (balanced) the risk for problems to an individual is similar to that with inversions: usually none if familial and slightly increased if de novo. Problems arise with translocations when gametes from a balanced parent are formed which do not contain both translocation products. When such a gamete combines with a normal gamete from the other parent the result is an unbalanced embryo which is partially monosomic for one chromosome and partially trisomic for the other.

Numerical and structural abnormalities can be further divided into two main categories: constitutional, those you are born with; and acquired, those that arise as secondary changes to other diseases such as cancer.

Sometimes individuals are found who have both normal and abnormal cell lines. These people are called mosaics and in the vast majority of these cases the abnormal cell line has a numerical chromosome abnormality. Structural mosaics are extremely rare. The degree to which an individual is clinically affected usually depends on the percentage of abnormal cells. A routine Cytogenetic analysis typically includes the examination of at least 15-20 cells in order to rule out any clinically significant mosaicism.

These are just some of the more common abnormalities encountered by a Cytogenetic Laboratory. Because the number of abnormal possibilities is almost infinite, a Cytogeneticist must be trained to detect and interpret virtually any chromosome abnormality that can occur.

## Examples of Chromosome Abnormalities

*Fig. 5.6* Inversion 3 Cell and Ideogram

*Fig. 5.7* Deletion 7 Karyotype

*Fig. 5.8* Dicentric (13; 14) Karyotype

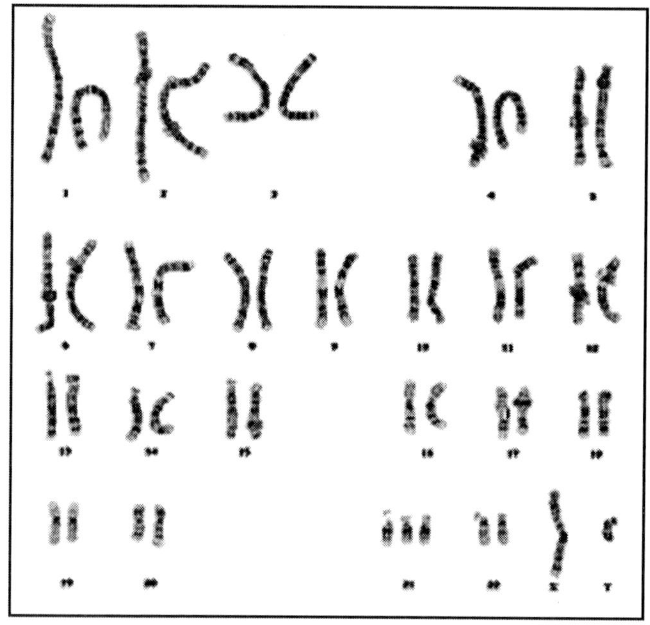

**Fig. 5.1:** Down Synrdome Karyotype

This karyotype is an example of Down Syndrome (trisomy 21), the most common numerical abnormality found in newborns. It is characterized by an extra chromosome 21 and the karyotype is written as: 47,XY,+21. The key to the karyotype description is as follows:

- 47: the total number of chromosomes (46 is normal).
- XY: the sex chromosomes (male).
- +21: designates the extra chromosome as a 21.

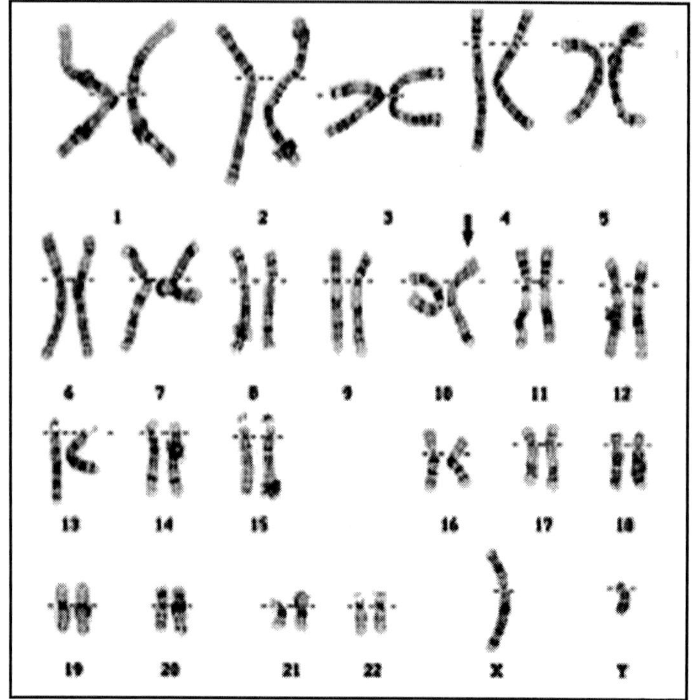


**Fig. 5.2:** Inversion 10 Karyotype

This karyotype is an example of an inversion, one of the more common structural rearrangements. In this case a segment in the q, or long arm of the right chromosome 10 is inverted. Since both breaks occurred in the long arm and the centromere is not involved, this is referred to as a paracentric inversion. If separate breaks had occurred in both the long and short arms the centromere would be inverted as well, this would be called a pericentric inversion. The karyotype is written as: 46,XY,inv(10)(q11.23q26.3). The key to the karyotype description is as follows:

- 46: the total number of chromosomes.
- XY: the sex chromosomes (male).
- inv(10): inversion in chromosome 10.
- (q11.23q26.3): break points of the inverted segment.

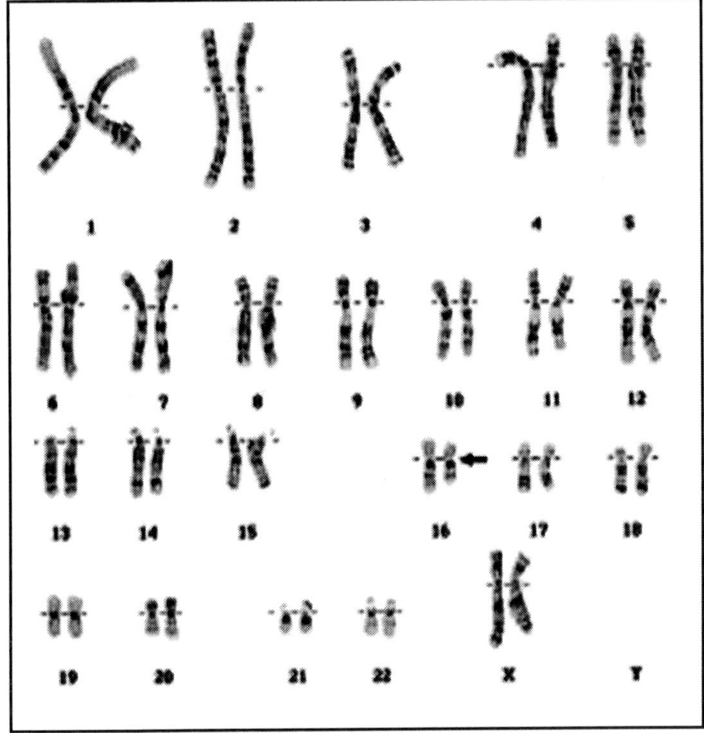


**Fig. 5.3:** Deletion 16 Karyotype

This karyotype is an example of a simple deletion in one chromosome. In this case a segment within the q, or long arm of the right chromosome 16 is deleted. In this particular example there are two microscopically visible breaks within the long arm making it an interstitial deletion. If there had been one break resulting in the loss of the end of a chromosome this would be called a terminal deletion. The karyotype is written as: 46,XX, del(16)(q13q22). The key to the karyotype description is as follows:

- 46: the total number of chromosomes.
- XX: the sex chromosomes (female).
- del(16): deletion in chromosome 16.
- (q13q22): break points of the deleted segment.

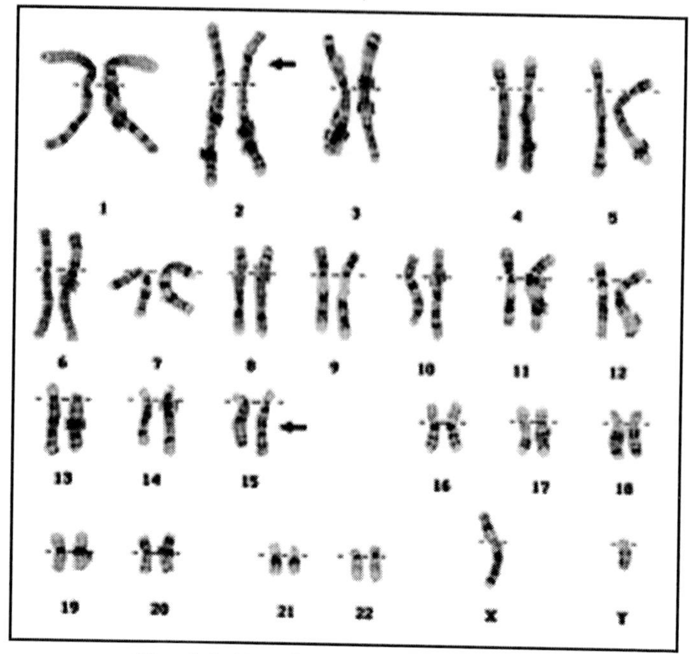


**Fig. 5.4:** Translocation (2;15) Karyotype

This karyotype is an example of a balanced translocation between two chromosomes. In this case a large segment in the *p*, or short arm of the right chromosome 2 has been exchanged with basically the entire *q*, or long arm of the right chromosome 15. Because the size of the exchanged segments is about equal, this particular structural rearrangement would be almost impossible to detect without banding techniques. The karyotype is written as: 46,XY,t(2;15) (p11.2;q11.2). The key to the karyotype description is as follows:

- 46: the total number of chromosomes.
- XY: the sex chromosomes (male).
- t(2;15): translocation between chromosomes 2 and 15.
- (p11.2;q11.2): break points in chromosomes 2 (p11.2), and 15 (q11.2) respectively.

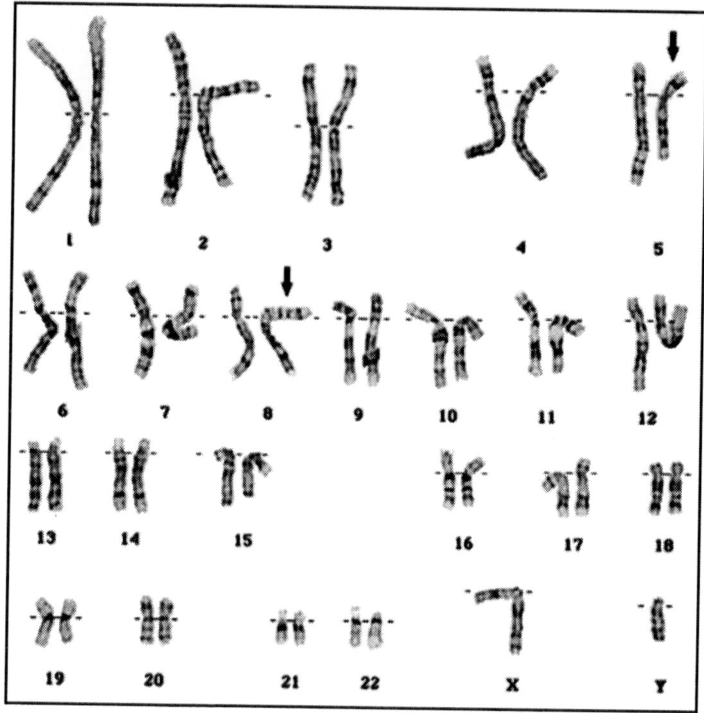


Fig. 5.5: Translocation (5; 8) Karyotype

This karyotype is an example of a balanced translocation between two chromosomes. In this case a large segment in the *q*, or long arm of the right chromosome 5 has been exchanged with a small segment from the *p*, or short arm of the right chromosome 8. The karyotype is written as: 46,XY,t(5;8)(q31.1;p23.1). The key to the karyotype description is as follows:

- 46: the total number of chromosomes.
- XY: the sex chromosomes (male).
- t(5; 8): translocation between chromosomes 5 and 8.
- (q31.1;p23.1): break points in chromosomes 5 (q31.1), and 8 (p23.1) respectively.

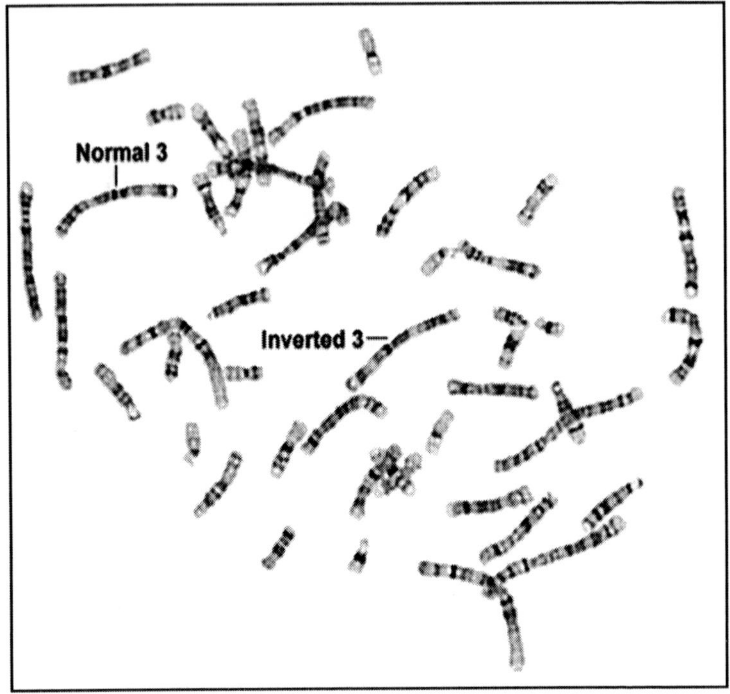


**Fig. 5.6:** Inversion 3 Cell and Ideogram

This metaphase cell is an example of a very subtle inversion. This is how the cell looks to a Cytogeneticist as it is viewed under a light microscope. In this case a segment in the *q*, or long arm of the indicated chromosome 3 is inverted. Since both breaks occurred in the long arm and the centromere is not involved, this is referred to as a paracentric inversion. The karyotype is written as: 46,XX,inv(3)(q24q27). The key to the karyotype description is as follows:

- 46: the total number of chromosomes.
- XX: the sex chromosomes (female).
- inv(3): inversion in chromosome 3.
- (q24q27): break points of the inverted segment.

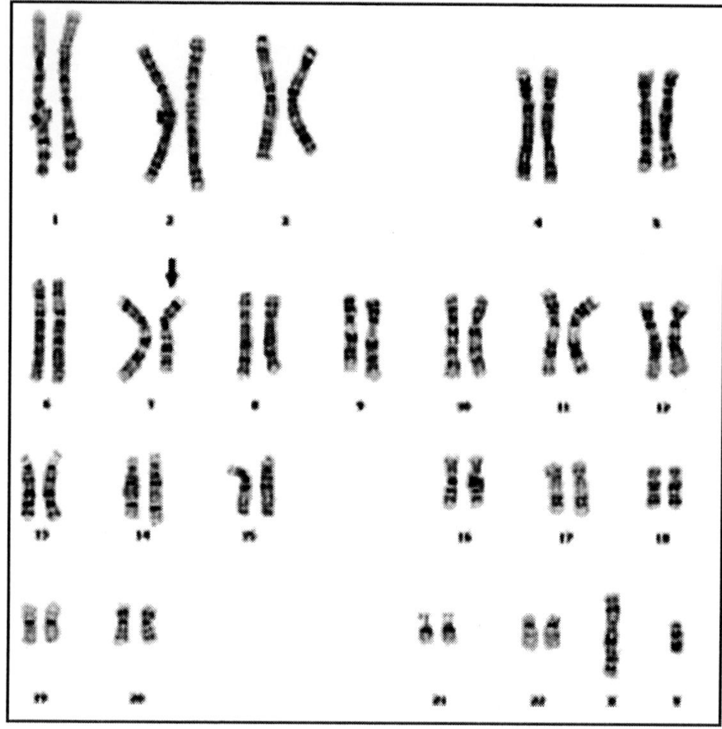

Fig. 5.7: Deletion 7 Karyotype

This karyotype is an example of a simple deletion in one chromosome. In this case a segment within the *q*, or long arm of the right chromosome 7 is deleted. In this particular example there are two microscopically visible breaks within the long arm making it an interstitial deletion. If there had been one break resulting in the loss of the end of a chromosome this would be called a terminal deletion. The karyotype is written as: 46,XY,del(7)(q11.23q21.2). The key to the karyotype description is as follows:

- 46: the total number of chromosomes.
- XY: the sex chromosomes (male).
- del(7): deletion in chromosome 7.
- (q11.23q21.2): break points of the deleted segment.

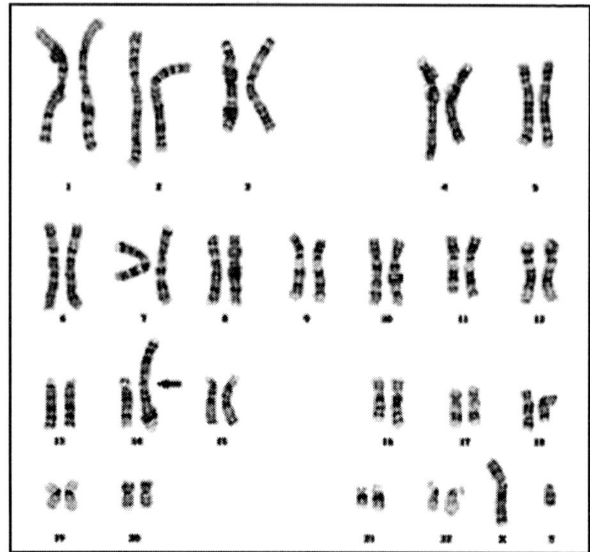

**Fig. 5.8:** Dicentric (13;14) Karyotype

This karyotype is an example of a special type of translocation involving the entire long arms, and quite often the centromeres, of acrocentric chromosomes. It is called a Robertsonian Translocation. In this case the entire *q*, or long arm, plus centromere of a chromosome 13 has been fused with the entire q arm, plus centromere of a chromosome 14. This particular example is unbalanced and results in trisomy 13. There are two normal chromosome 13's plus the chromosome 13 involved with the translocation, thus there are three copies of chromosome 13. The translocation is shown as the right chromosome 14 in the karyotype. The karyotype is written as: 46, XY, + 13, dic(13;14)(p11.2;p11.2). The key to the karyotype description is as follows:

- 46: the total number of chromosomes. The chromosome number remains 46 because the long arms of chromosomes 13 and 14 have basically fused into one chromosome.
- XY: the sex chromosomes (male).
- +13: indicates the presence of an extra chromosome 13.
- dic(13;14): dicentric chromosome involving chromosomes 13 and 14. As with many Robertsonian translocations, the centromeres of both chromosomes are present, thus the "dicentric" designation.
- (p11.2;p11.2): breakpoints in chromosomes 13 (p11.2), and 14 (p11.2) respectively.

# 6

# Fluorescent *In Situ* Hybridisation of Plant Chromosomes

## *Illuminating the Musa Genome*

### Introduction

Characterisation of banana and plantain germplasm has until now, been largely based on the use of phenotypic characters and more recently on molecular markers such as RFLP and RAPD. Cytogenetic studies have proved difficult in the genus Musa because of the small size of the genome, just 10 per cent of the barley genome for example, and the large number of chromosomes (2n=3x=33 in most banana cultivars, compared to 2n=2x=14 in barley). Molecular cytogenetic studies, which link data about the molecular composition and organisation of the genome with the chromosomes, offer greater understanding of phylogenetic relationships and improved clarity of taxonomic discrimination, allowing the identification of aneuploids and assisting selection.

In recent years, there have been rapid advances in the direct observation and analysis of banana chromosomes using molecular cytogenetic methods. This focus paper provides some information on the applications of such techniques in relation to banana and plantain research.

## *In Situ* Hybridisation

The *in situ* hybridisation (ISH) technique, developed more than 30 years ago (Gall and Pardue 1969, John et al. 1969) allows genes or DNA sequences to be directly localised on chromosomes in cytological preparations. The development of user-friendly fluorescent techniques (Langer-Safer et al. 1982, Pinkel et al. 1986) has greatly increased the application of this technique during the last 15 years. Fluorescent *in situ* hybridisation (FISH) allows hybridisation sites to be visualised directly and moreover, several probes can be simultaneously detected with different fluorochrome, allowing the physical order on the chromosomes to be determined.

For the FISH technique, DNA sequences to be localised are first labelled to produce the probe.

The probe is coated on the target chromosome which is spread in a hybridisation buffer. After treatment to denature the DNA into single strands, the probe and target are allowed to re-anneal. The probe will bind specifically to the complementary site on the chromosome. After washing and detection with a fluorescent reporter, a discrete fluorescent signal is visible at the site of probe hybridisation, which can be visualised using a fluorescent microscope (Figure 6.1).

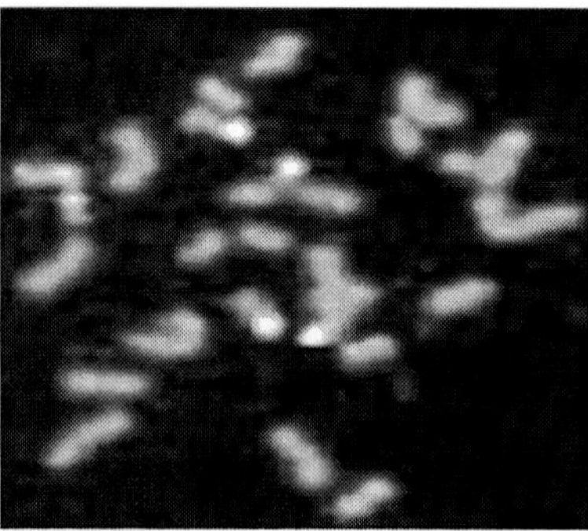

**Fig. 6.1:** Double FISH showing the rDNA sites on somatic metaphase chromosomes of Narenga. The 18-25S rDNA site are visualised in green (FITC) and the 5S rDNA sites are visualised in red (Texas Red). The chromosomes are counterstained with DAPI (blue). (*Courtesy of CIRAD*)

One of the important modifications of the ISH technique is genomic *in situ* hybridisation (GISH). GISH is a genomic painting technique which allows parental genomes in interspecific hybrids to be distinguished (Figure 6.2). Total genomic DNA from one parent is labelled as a probe and unlabelled total DNA of the other parent is used as a block. Alternatively, total DNA from both parents is labelled and these are both used as probes, each one revealed with a different fluorochrome. This technique is based on the rapid evolution during speciation of repeated sequences, which represent the major part of plant DNA. If the species are distant enough, the repeat sequences allow the chromosomes from the two parental species to be differentiated.

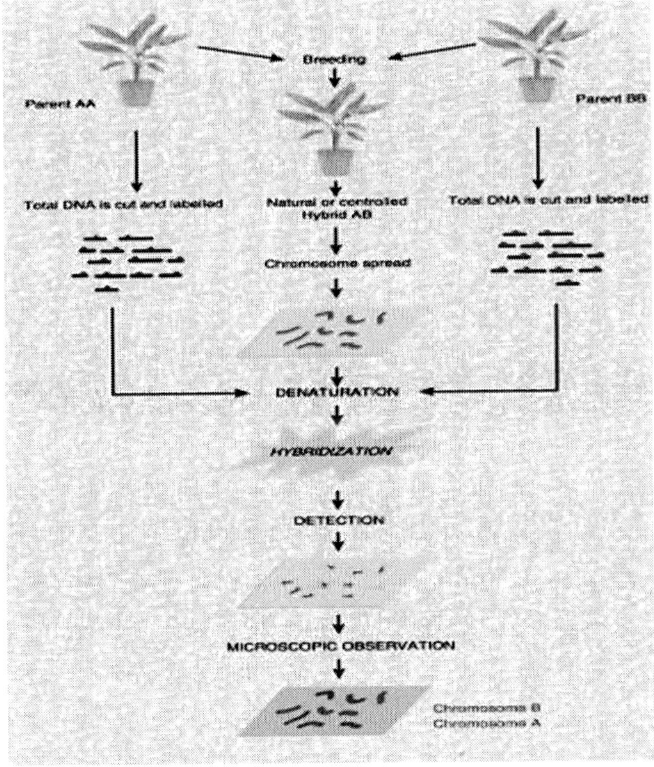


**Fig. 6.2:** Principle of genomic *in situ* hybridisation. (*Courtesy of CIRAD*)

## Applications

Untangling the A, B, S and T genomes by genomic *in situ* hybridisation The classification of Musa cultivars into genomic groups has so far been based on chromosome numbers and morphological traits (Cheesmann 1947, Simmonds and Shepherd 1955) as well as more recently, on molecular markers. GISH, however, provides a powerful complementary tool to molecular markers, enabling the portion of the genome contributed by each parental species in interspecific hybrids and their derivatives to be visualised (Figure 6.3). This technique has allowed the chromosomes from the four wild Musa species, M. acuminata, M. balbisiana, M. schizocarpa and the Australimusa species, involved in the origins of cultivated bananas to be differentiated.

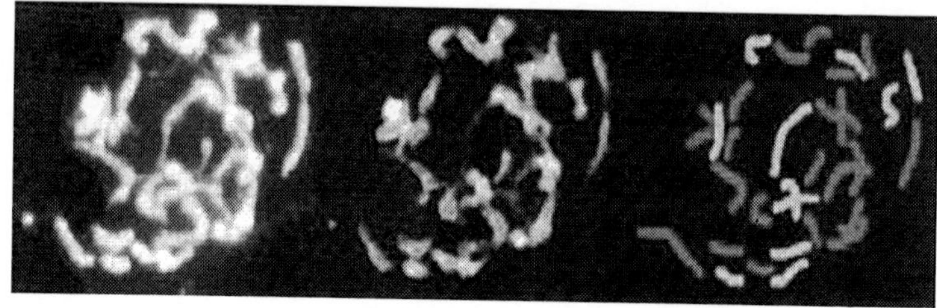

**Fig. 6.3:** Metaphase oftriploid plantain Mbi Egome (AAB): *(a)* The 33 chromosomes stained blue with the DNA stain DAPI; *(b)* In situ hybridisation f genomic A DNA (red) and B genomic DNA (green); *(c)* Interpretation shows the d labelled regions on 22 chromosomes; the other 11 chromosomes are labelled only with green. (*Courtesy of John Innes Center*)

The exact genome structure of several interspecific cultivars has been examined using GISH. The results were in most cases consistent with the chromosome constitution estimated through phenotypic descriptors, with one notable exception. The clone 'Pelipita', was found to contain 8A and 25 B chromosomes, instead of the 11A and 22 B predicted (Figure 6.4).

Using molecular markers, it was recently confirmed that the species M. schizocarpa (S genome) and species of the Australimusa section (T genome) have contributed to the origin of some cultivars. However, it was not possible to determine what proportion of these species are present in the genome. Using GISH

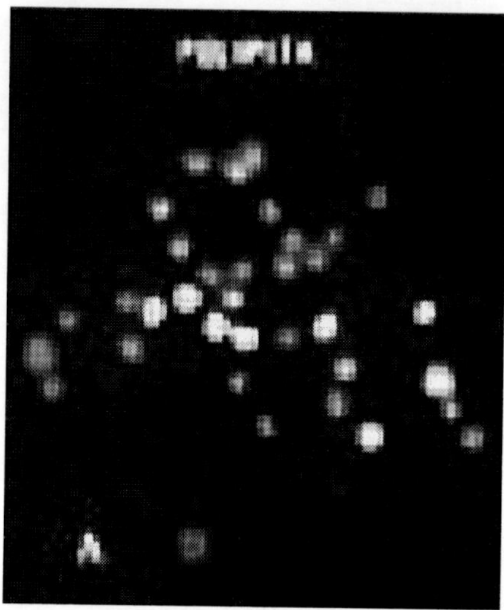

**Fig. 6.4:** GISH on somatic metaphase chromosomes of 'Pelipita'using total NA from a BB clone revealed in red with Texas Red and total DNA from an AA clone revealed in green with FITC. (*Courtesy of CIRAD*)

it was possible to demonstrate for example, that the S genome contributed a full set of S chromosomes to the cultivar Wompa. Similarly, GISH showed that one basic set of T chromosomes are present in the cultivars 'Karoina'and 'Yawa 2'and established their genome constitution as AAT and ABBT, respectively (Figure 6.5).

Identifying individual Musa chromosomes and visualising DNA sequences Individual chromosomes are difficult to identify conventionally because they are so similar.

However, individual chromosomes can be defined by the hybridisation of specific cloned or synthetic repetitive DNA sequences. For example the 18S-25S rDNA is present at a single site in each genome and can be used to define that chromosome. This has a further significant use as this single site in each genome enables easy assessment of basic ploidy levels in hybrid or tissue culture material.

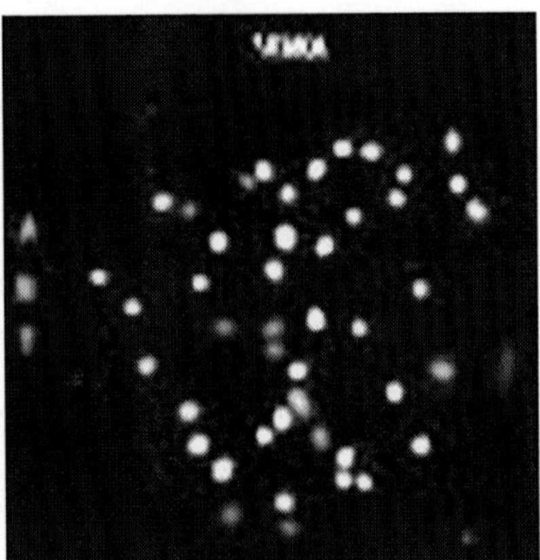

**Fig. 6.5:** GISH on somatic metaphase chromosomes of 'Yawa 2' using total DNA from a AA clone revealed in green with FITC, total DNA from a BB clone revealed in red with Texas Red and DAPI counterstaining (blue). (*Courtesy of CIRAD*)

The hybridisation pattern obtained can also provide indicators of recent and evolutionary rearrangements in the genomes (Figure 6.6).

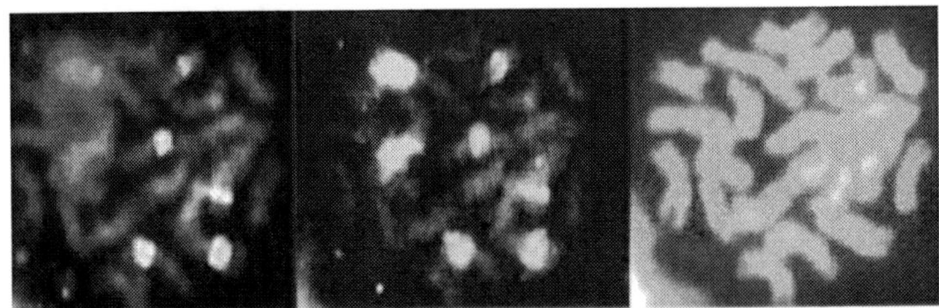

**Fig. 6.6:** *In situ* hybridisation to chromosomes of an AA Musa hybrid: *(a)* The 22 chromosomes stained blue with DNA stain DAPI; *(b)* Five sites of hybridisation to 5S rDNA probe (green); *(c)* Single site hybridisation to 18S-25S rDNA probe on each of the two genomes. (*Courtesy of John Innes Center*)

The development of similar markers (repeated sequences, BAC, etc.) for the various linkage groups will enable the different chromosomes to be assigned to respective linkage groups and will thus efficiently complement genetic mapping efforts. This would also open the way for the investigation of structural re-arrangements which are reported to be frequent in bananas.

These re-arrangements result in important irregularities in meiosis and irregular chromosome transmission and may have been involved in the development of sterility, a prerequisite for edible fruit.

## Understanding BSV

FISH can be used to analyse the numbers and loci of other chromosomal sequences and it has been used to analyse the integration of banana streak virus (BSV) DNA into the Musa genome. Numerous lines of evidence including PCR and genomic.

Southern analysis pointed to the possible integration of BSV sequences. To examine whether these BSV sequences in high molecular weight DNA were actually in the Musa nuclear chromosomes, double target *in situ* hybridisation was conducted on chromosomes from the plantain cultivar Obino L'Ewai, using a probe specific to BSV and a probe specific to a Musa sequence. Both probes gave hybridisation signals on chromosomes of Obino L'Ewai. A major hybridisation site to BSV was detected on both chromatids of one chromosome in each metaphase and at least one weaker hybridisation site was regularly seen. This clearly demonstrates that viral sequences are integrated in the nuclear genome.

The Musa probe showed hybridisation to multiple sites throughout the genome, including near the major BSV site, but was not uniformly dispersed.

Representatives of AA, AAA and BB genome Musa were analysed by FISH and all showed clear hybridisation of BSV sequences. The strength of the signals indicates that multiple copies of the target sequence were integrated at most of the observed sites (Figure 6.7.) This is further compelling evidence that BSV sequences are integrated into the Musa genome and that this integration must have been an ancient event.

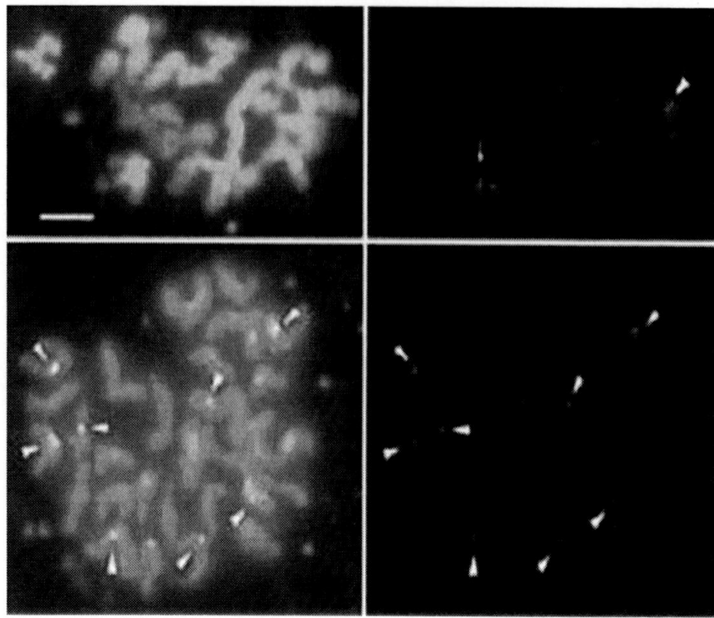

**Fig. 6.7:** Musa genotypes Cavendish (AAA) and Obino L'Ewai (AAB) showing hybridising (integrated) BSV sequences. *In situ* hybridisation to metaphase spreads of Obino L'Ewai: *(a)* The 33 chromosomes stained blue with the DNA stain DAPI. *(b)* Hybridisation sites of BSV (red) showing one major site in each metaphase (arrowhead) and at least one minor site (arrow). *In situ* hybridisation to metaphase spreads of Dward Cavendish: *(c)* The 33 chromosomes stained blue with the DNA stain DAPI. *(d)* Hybridisation sites of BSV (red) showing at least eight ajor site in each metaphase (arrowhead). (*Courtesy of John nnes Center*) Bar = 5 mm

## Visualisation of Fine Scale DNA Structure

The organisation of gene and DNA structures can be visualised by a relatively new method, that of *in situ* hybridisation of probes to DNA fibres extended to their full molecular length. Theoretical considerations of the length of the extended DNA molecule and calibration from hybridisation with probes of known length and interspersion pattern can relate the lengths of observed fibres to the numbers of bases (Figures 6.8 and 6.9).

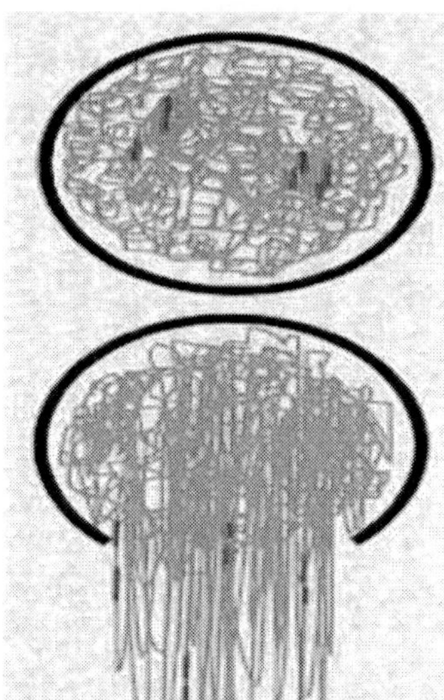

**Fig. 6.8:** Cartoon of an interphase nucleus fixed to a slide: *(a)* The blue chromatin labelled at eight sites by a red *in situ* hybridisation probe; . After lysis of the nucleus and tilting of the slide the DNA fibres are stretched to their full molecular length. They can hybridise with the ame probe and now clearly show the relationship between probe and fibre. (*Courtesy of John Innes Center*)

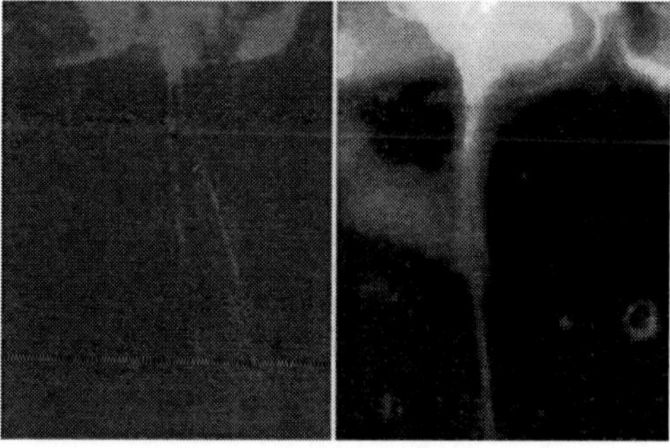

**Fig. 6.9:** Rye interphase nucleus: *(a)* DAPI staining shows the strechted DNA as blue fibres running downwards. *(b)* A highly repetitive ibosomal DNA probe labels multiple sites on some but not all of the fibres. Here, the fibres are too bundled for detailed analysis of the gene structure, but the relationship between nucleus, the fibres and the probe is evident. (*Courtesy of John Innes Center*)

This technique was used to investigate the structure of the integrated BSV sequence. A genomic clone (Ndowora et al. in press) and PCR based methods (Harper et al. in press) had shown that the integrated sequence adjacent to a Musa interspersed sequence was complex, containing an inverted region and some very highly rearranged stretches. Stretched DNA fibres were prepared on slides from Obino L'Ewai nuclei. Double-target hybridisation with the genomic Musa sequence and BSV showed long rows of hybridisation sites ('dots') along stretched DNA fibres. The Musa sequence was present at sites associated with the BSV hybridisation sites and also independently as variable lengths of rows of dots (Figure 6.10). It was apparent that there were two different lengths of Musa-BSV chains of dots present in approximately equal numbers. Some were 50 μm long, representing structures containing multiple copies of BSV sequences (150 kb long) and others were 17 μm long (about 50 kb structures). Each group of fibres, long and short, showed common patterns of red and green signal sites and gaps, with repeating units of BSV sequence adjacent to Musa sequence.

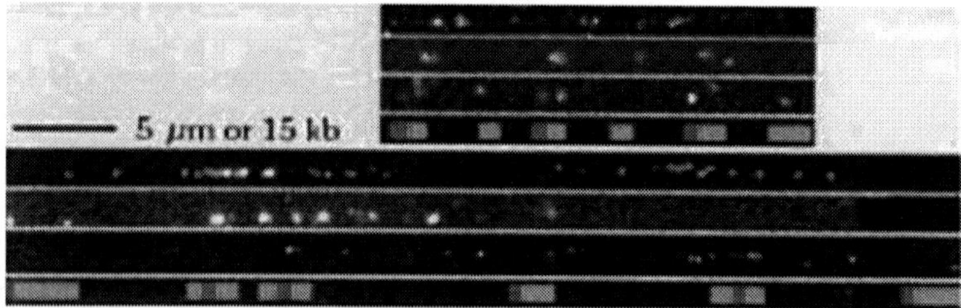


**Fig. 6.10:** Fibre stretches of Musa Obino L'Ewai AAB showing hybridising BSV and associated Musa sequences. *In situ* hybridisation to extended DNA fibres from Obino L'Ewai nuclei. Green and red dots represent probe hybridisation sites to BSV sequences and associated Musa sequences respectively. Two different patterns of chains of dots were detected: *(a)* Three independent and aligned long fibres above a consensus diagram of hybridisation pattern showing red sites and chains of green signals. Both the Musa and BSV sequences are present in multiple copies in the structure of 150 kb, in at least two different relative orientations, and are separated by gaps with no hybridisation (no homology to probes). *(b)* Three aligned short fibres above consensus diagram, showing a pattern that can be interpreted as three sub-repeats. Under the hybridisation, detection and imaging procedures used, individual signals are larger than expected from the probe length, may be slightly displaced from the axis, and some supposed target sites may not have a detectable signal. (*Courtesy of John Innes Center*) Bar = 5 m, corresponding the 15 kb DNA fibre length.

The longer structure is considered to correspond to the major hybridising site seen on metaphase chromosomes while the shorter structure, corresponded to the minor hybridisation site.

## Conclusion

Molecular cytogenetic methods are adding a powerful set of tools to those already available to study genome organisation, evolution and recombination. GISH has immense potential for identification of chromosome origin and can be used to characterise cultivars and hybrids produced by Musa breeding programmes.

Repetitive and single copy DNA probes are yielding insights into the relationship between genetic and physical maps of Musa and genome evolution. Finally, fibre *in situ* hybridisation can be used to examine the organisation of genes and DNA sequences. Together, these techniques provide data for Musa breeders, which can be used to tackle the challenges caused by banana streak virus, tissue culture and somaclonal variation, the use of wild germplasm in breeding and the irregular transmission of chromosomes during meiosis. *In situ* hybridisation therefore holds great potential to help scientists develop optimum breeding strategies in order to create high quality and disease resistant bananas.

# 7

# PD-loop Technology

## Introduction

DNA labelling with different probes is the traditional way to detect and manipulate a specific sequence in DNA samples. Usually, single-stranded (ss)DNA — obtained in most cases by denaturation — is employed for these purposes to be recognised by DNA/RNA probes or their synthetic analogs via Watson-Crick pairing. In contrast tossDNA, double-stranded (ds)DNA, the most common DNA structure, lacks such a simple and sequence-universal recognition principle. This poses a general problem of sequence-specific detection and handling of dsDNA.

In practical terms, the problem is: how to selectively target fragments of non-supercoiled dsDNA, which is the typical form of DNA in a majority of analytical samples? Several approaches have been developed in this direction. Formation of inter-molecular triplexes between the dsDNA target and a probe via Hoogsteen pairing was first proposed. Notwithstanding the value of triplex-targeting strategy in some applications , it is limited by mostly homopurinehomopyrimidine regions. Recognition of mixed-base dsDNA sequences by oligonucleotide probes can be facilitated by RecA protein or its fragment. However, the fidelity of RecA-assisted DNA recognition is much lower than that of 'pure' DNA-DNA or DNA-RNA interactions, being tolerable up to four mismatches.

Sequence-specific binding of long ssDNA molecules or a pair of 'pseudocomplementary' oligonucleotides to dsDNA targets containing mixed sequence of nucleobases via strand invasion has been demonstrated. However, these protein-free complexes were formed only at the end of DNA duplexes. Rloops may be formed inside linear dsDNA, but partial DNA denaturation and long RNAs are necessary . T hus, none of the approaches involving DNA or RNA type probes offer a fully satisfactory solution to the aforementioned problem.

The invention of the nonionic DNA/RNA mimic, peptide nucleic acid (PNA; Fig 7.1) , has launched the development of new approaches in this area. Several types of PNA exhibiting strand-invasion abilities have been described, with pyrimidine PNAs and mixed-base pseudocomplementary PNAs (pcPNAs) being potentially most practical.

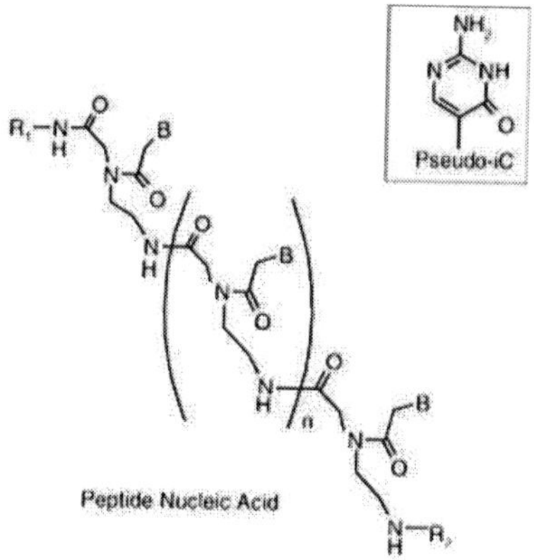

Pseudo-iC

Peptide Nucleic Acid

**Fig. 7.1:** Generic chemical structure of common PNAs with 'classical' achiral and uncharged N-(2-aminoethyl)glycine periodic backbone, which is homomorphous to the sugar-phosphate DNA/RNA backbones and to which ordinary or modified DNA/RNA nitrogenous bases (shown here as B) are linked. In case of bis-PNA formed by a pair of the flexibly linked pyrimidine PNAs, [] all cytosines in one PNA oligomer are replaced by pseudoisocytosines (pseudo-iC or J bases; shown as inset) to avoid the sharppH dependence of the (PNA)2-DNA invasion triplex formation ]. In pcPNAs ], adenines are replaced by diaminopurines, whereas thymines are replaced by thiouracils, thus allowing the formation of PNA-DNA doubleduplex invasion complexes. R1 and R2 represent cationic lysine residues used to increase the PNA solubility and binding affinity and bis-PNA repetitive linker units (usually dubbed eg1) based on 8-amino-3,6-dioxaoctanoic acid.

The use of PNA-based probes makes it now possible to more effectively target internally located dsDNA sites. Despite many advantages these new probes are offering, they have their own drawbacks. First, binding of pyrimidine PNAs is again limited to homopurine-homopyrimidine DNA target sites. Second, stable PNA-DNA strand-invasion complexes are formed even with rather short (8–10 bp) DNA sites, which occur relatively frequently and therefore are not unique in long DNAs. Third, PNA oligomers and PNA-DNA or PNA-RNA heteroduplexes are mostly inert with regard to enzymatic transformations. This feature limits those potential applications that require further processing of hybridised probes.

Recently, my colleagues and I have proposed another approach for binding a probe to dsDNA, which is based on the PNA-assisted assembly of a stable complex between dsDNA and an oligonucleotide or other probes with mixed purine-pyrimidine sequence in a protein-free system. The idea proved to be very fruitful and laid the foundation of PD-loop technology, which employs the formation of so-called PD-loops and related structures.

## What is the PD-loop?

All assays described here involve the successive PNA-assisted formation of locally looped DNA structures: first P-loops, also known as (PNA)2-DNA invasion triplexes, then PD-loops (i.e., the PNA-distended DNA loops). Within P-loops, two pyrimidine PNA oligomers, which are commonly linked into bis-PNA to enhance the strand-invasion efficiency, form a triplex with the complementary purine site on one DNA strand, leaving the other DNA strand displaced (*Fig. 7.2A*). When two Ploops are located close to each other, they merge, yielding an extended P-loop (*Fig. 7.2B, C*). In such a case, a larger singlestranded region occurs inside dsDNA, serving as a selective target for binding of an oligodeoxynucleotide probe containing all four nucleobases via Watson-Crick pairing (*Fig. 7.2D*). As a result, the PD-loop appears, which consists of locally open dsDNA, a pair of bis-PNA 'openers' and an oligodeoxyribonucleotide. Similar complexes, PR- and PP-loops, can be assembled, in which an oligodeoxynucleotide probe is substituted by an oligoribonucleotide or PNA oligomer, respectively.

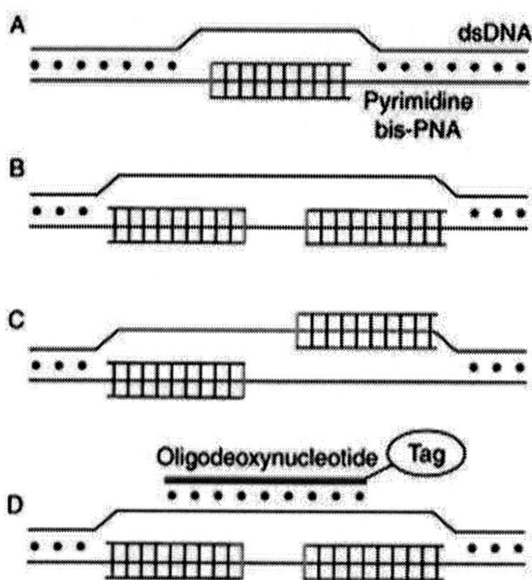


**Fig. 7.2:** Looped structures formed within dsDNA through PNA invasion: single P-loop (A), merged or extended P-loop (B,C) and PD-loop (D). Besides Watson-Crick and Hoogsteen base pairing, the (PNA)2-DNA triplexes formed by pyrimidine bis-PNAs are additionally stabilised by a complex net of bonds, water bridges and van der Waals interactions between DNA and PNA backbones. The oligonucleotide probe can be radioactively labelled and/or may carry a special tag like biotin, fluorescein etc. Arrangement of PNA openers depicted in (C) was used for pre-gel hybridisation of a 32P-labelled oligonucleotide probe to the 9.4 kb fragment of Hind III-digested DNA from bacteriophage ? (*Unpublished Data By DEMIDOV VV*).

The PD-loop formation is an exceptionally sequence-specific process because most of the mismatched sites will not be exposed by PNA openers. In fact, only those incorrect targets could readily be opened, which contain a mismatch between the PNA openers binding sites and hence in the middle of the probe recognition sequence. Considering the small length of the oligonucleotide probe required for the PD-loop formation, usually 15 nt or less, such a mismatch will be very unfavourable, thus making the mismatched oligonucleotide-DNA complexes unstable. Furthermore, a typical PD-loop spans about 20 bp or more so it is normally unique in any given genome. Consequently, the PD-loop construction makes it possible to selectively target the designated site within long DNAs.

In view of possible diagnostic applications, the sequence limitations on the PD-loop formation due to the homopyrimidine nature of PNA openers are rather mild. Indeed, our data and results of others demonstrate that homopurine binding sites for PNA openers can be very short, down to 5 bp. Furthermore, they can be located on the same or on opposite DNA strands and be separated by an arbitrary sequence of nucleobases up to 10 bp long. Statistically, DNA sites with sequences that meet these requirements should occur quite frequently, i.e every several hundred base pairs of a random DNA sequence, on average. This estimation is supported by our analysis of genomic sequences now available for several prokaryotic and eukaryotic organisms. Thus, normally each gene must carry such a site and any genome will contain many of them.

The use of PD-loops for sequence-specific handling of duplex DNA is demonstrated here by four recently developed applications:

- dsDNA affinity capture
- Topological labelling
- Non-denaturing sequencing
- Fluorescent detection

Examples of PD-loop technology

Duplex DNA Capture

The procedure, known as oligonucleotide/PNA-assisted affinity capture (OPAC), involves a three-step protocol . First, two cationic bis-PNAs bind the designated DNA duplex in a mixture of dsDNA restriction fragments or from some DNA library etc. and locally separate the target site strands. Then, a complementary biotinylated oligonucleotide hybridises to the exposed DNA single strand forming a PD-loop complex. Finally, the capture of the chosen dsDNA and its subsequent release is performed using streptavidin-covered iron microbeads and magnetic separator.

Release of captured dsDNA from the microbeads can be performed by incubation of the sample in a low-salt solution (usually, neutral Tris-buffer containing 50 mM NaCl) at 45-50°C for 20 min. The temperature of the elution will depend on the binding affinity of an oligonucleotide probe, which is determined by salt concentration and oligonucleotide length and base composition. Under these conditions for dsDNA elution, the PNA openers remain bound to dsDNA. Thus, up to five rounds of OPAC selection can be performed without the time-consuming retargeting of DNA samples with the PNA openers. Only the oligonucleotide probe needs to be rebound to DNA target to facilitate the next round of OPAC enrichment. If a single round of OPAC is enough for dsDNA selection or when it is necessary to remove bound PNA from DNA for subsequent manipulation with captured materials, high-salt elution buffer (e.g., Tris-buffer containing 1 M NaCl) should be used at the end of the procedure. In this case, the PNA-free DNA target can be released by incubation of the sample at 65°C for 20 min.

Fig. 7.3B shows the results of the multiround OPAC procedure used for isolation of a specific, about 1 kb-long dsDNA fragment from a restriction digest of the entire yeast genomic DNA. One can see that after three rounds of OPAC selection, the complexity of the initial DNA pool (~ 40,000 different DNA fragments) was greatly reduced. Additional rounds of selection yielded virtually pure target DNA fragment (indicated by arrow), whose identity and homogeneity was confirmed by sequencing and restriction analyses.

Our data demonstrate that the OPAC method has the potential to quantitatively capture larger, up to 10 kb-long dsDNA fragments (*Unpublished Data By Demidov vv, izvolsky ki*). Thus, the OPAC methodology makes it possible to purify gene-sized chunks of duplex DNA and hence isolate any gene of interest.

This situation is radically different from the case of the tri lex affinity capture. Normally, this method requires a homopurine- homopyrimidine target region to be at least 20 bp-long to provide a quantitative isolation of the DNA duplex. Such

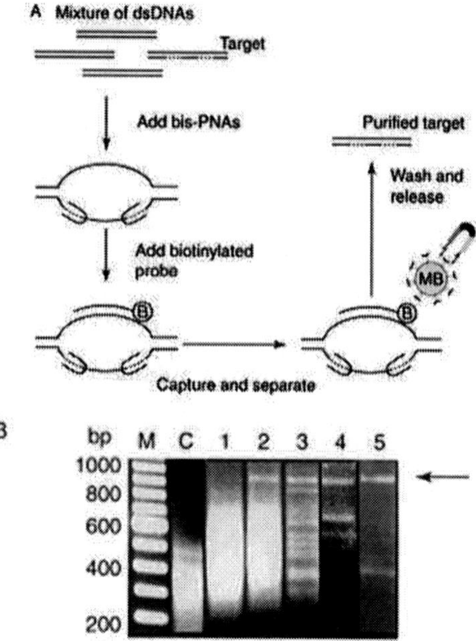


**Fig. 7.3:** Duplex DNA Figure capture. (A) Schematics illustrating the key steps of the biomagnetic OPAC procedure. B: Biotin; MB: Streptavidin-covered iron microbead. (B) The OPAC purification of designated fragment from entire yeast DNA (S. cerevisiae strain AB1380) treated with the Mse I restriction enzyme. The PD-loop-forming site, 5'A3GA2G2CTG2A2G2A33' (here and below, Nn means that a nucleobase N, or similarly Lys and PNA linker unit eg1, is repeated n times at a specific position of a sequence), is located on the 903 bp-long fragment of Mse I-digested yeast chromosome IX and consists of two homopurine targets for PNA openers (underlined) separated by 4 bp of mixed purine-pyrimidine sequence. This unique 19 bp site was targeted at the beginning of the first round with the bis-PNAs H(Lys)2-T3JT2J2-(eg1)3- C2T2CT3-LysNH2 and H(Lys)3-T2J2T3-(eg1)3-T3C2T2-LysNH2. Affinity capture was performed with the use of biotinylated 14-mer oligonucleotide biotin- 5'GA2G2CTG2A2G2A3'. For experimental details, see refs. [25,30]. Lanes 1–5 correspond to one, two, three, four and five rounds of PD-loop formation/ biomagnetic separation procedure, respectively. Lane M represents a 100 bp DNA ladder; lane C is the initial pool of yeast DNA fragments. An aliquot of captured DNA collected after each round of procedure was uniformly PCR amplified to analyse the enrichment using agarose gel electrophoresis.

a site occurs extremely rarely, statistically once per every million base pairs, hence the required target is normally not met in every gene. The same is true for the PNA-mediated affinity capture of long dsDNA stretches via the trinucleotide repeats. I therefore expect that the OPAC procedure can be used in different DNA technologies and DNA diagnostics for efficient isolation and purification of intact dsDNA.

## Topological DNA Labelling

By comparison with conventional hybridisation approaches employing linear probes, DNA labelling with circularised probes has two advantages. First, the use of circular oligonucleotides as labels offers significant improvement for DNA detection through the ability to perform rolling-circle amplification (RCA), which is also referred to as rolling-circle replication (RCR) assay. Second, catenation of a probe with the DNA target yields more stable and hence more localised labelling of target molecules.

Still, with the exception of our approach, which will be described here, all other approaches developed for DNA labelling with circularised probes have a common drawback. Specifically, they only provide a pseudotopological linkage, thus allowing the circular label to freely move along the targeted DNA. Such an unrestricted sliding of a circularised probe significantly compromises the localised detection of the target sequence, unless the additional ligand-modulated locking is used. Furthermore, dsDNA labelling approaches designed by others require long triplex-forming sequences, which are very rare. The PD-loop provides one more opportunity to assemble a catenated, rotaxane-like structure between the dsDNA target and oligonucleotide probe (*Fig. 7.4A*). Besides diminished sequence restrictions, assembly results in a true topological linkage of a circularized probe with dsDNA. In the final structure which resembles an earring, a segment of the circular label is threaded sequence-specifically between complementary strands of dsDNA. Therefore, in this construct, an oligonucleotide probe is immovably fastened to the dsDNA target.

*Fig. 7.4B* demonstrates the topological labelling of dsDNA target fragment carrying an artificial PD-loop-forming site. The DNA ligase treatment of the circularizable biotinylated oligonucleotide after its hybridisation to the pre-exposed dsDNA target results in the probe closing. This manifests in the formation of a low-mobility complex: the DNA duplex tagged with the earring probe (*Fig. 7.4B*). The streptavidin- induced extra retardation confirms that the biotin-labelled oligonucleotide probe does participate in the formation of the final pseudorotaxane complex (*Fig. 7.4B*). Resistance of our assembly against ExoVII nuclease, which digests linear but not circular oligonucleotides, and stability of the final complex to high temperature incubation, which dissociates even the PNA openers (not shown here), prove the probe circularisation and its immobility.

The study of the assembly of earring complexes performed with another PD-loop-forming site (*a part of the nef gene coding region of the HIV-1 virus; see legend to Fig. 7.6 for the sequence*) and some of its single-mismatched variants has demonstrated an extremely high sequence specificity of topological DNA labelling. We also showed that earring complexes can be assembled on closed circular plasmid DNA and on long phage DNA embedded in agarose gel and that the RCA reaction can be performed on earring probes by some DNA polymerases similar to padlock probes hybridised to long ssDNA. These data suggest that highly localized RCA-amplified dsDNA detection and other manipulations with DNA duplexes, including nanodesigns of higher order DNA assemblies, become possible through precise spatial positioning of circular oligonucleotides on the DNA scaffold.

## Non-denaturing dsDNA Sequencing

The PD-loop can function like a primosome: addition of DNA polymerase capable of strand-displacement activity to the PDloop initiates the primer-extension reaction directly within linear dsDNA. As it is shown schematically in *Fig. 7.5*, the combination of this approach with isothermal dideoxy chain termination protocol makes it possible to generate a sequencing ladder of several hundred nucleotides without denaturation of duplex DNA. *Fig. 7.6* shows that a highquality sequence read could be isothermally obtained via the PD-loop even for long dsDNA templates.

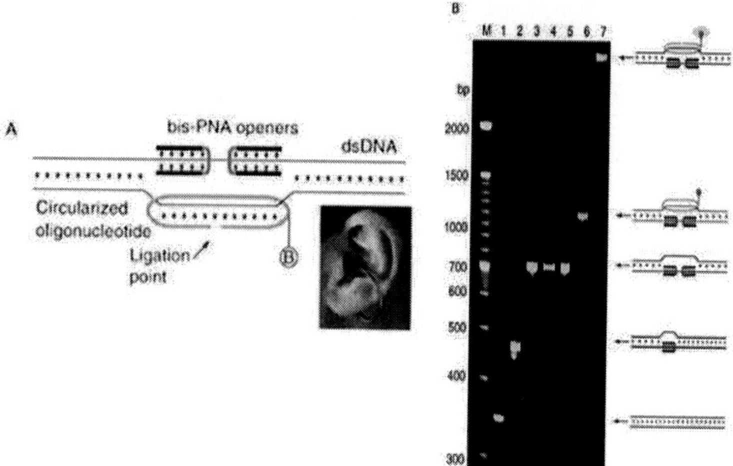


**Fig. 7.4:** Topological DNA labelling through the PD-loop formation. (A) Design of a topological label resembling an earring (B: Biotin; see inset for resemblance). In the hybridised state, the mini of circularisable earring probe precursor oligonucleotide are juxtaposed and can be enzymatically ligated. The typical rring probe makes one-two turns around the displaced DNA strand, thus linking itself to the dsDNA target. (B) Assembly of earring probes on the 350 bp target fragment of double-anded plasmid carrying the cloned artificial PD-loop-forming site 5'TC4T2CGA2C2T2CT33', where the complementary PNA binding sites are emboldened. PNA openers: H-TC4T2C-eg1-(Lys)2-eg1-JT2J4T-LysNH2 and HLys2-T3JT2J2-(eg1)3-C2T2CT3-LysNH2. 5'-phosphorylated circularisable oligonucleotide obe: 5'GA2G4AT3GT3CT2AXT2GT3AT3A2GA2G2T2C3', where X denotes a biotinylated reporter segment incorporated into the backbone and the 8- and 10-nt-long end stretches complementary to the DNA target are italicised. Formation of earrings was monitored by gel-shift assay. Lane M: 100 bp ladder; lane 1: DNA target fragment; lanes 2 and 3: Target DNA fragment incubated with only one or with both bis-PNA openers, respectively; lane 4 (negative control): Target fragment with both PNA openers incubated with T4 DNA ligase without oligonucleotide probe; lane 5 (negative control): Target fragment with both PNA openers and oligonucleotide probe without DNA ligase; lane 6 (formation of earrings): Target DNA fragment with both PNA peners and oligonucleotide probe incubated with DNA ligase; lane 7 (positive control): The earring complex analysed in lane 6 after addition of streptavidin shown as a large globule in the op right schematics; the binding of streptavidin to the ringconjugated biotin results in an additional retardation of the labelled DNA target fragment.

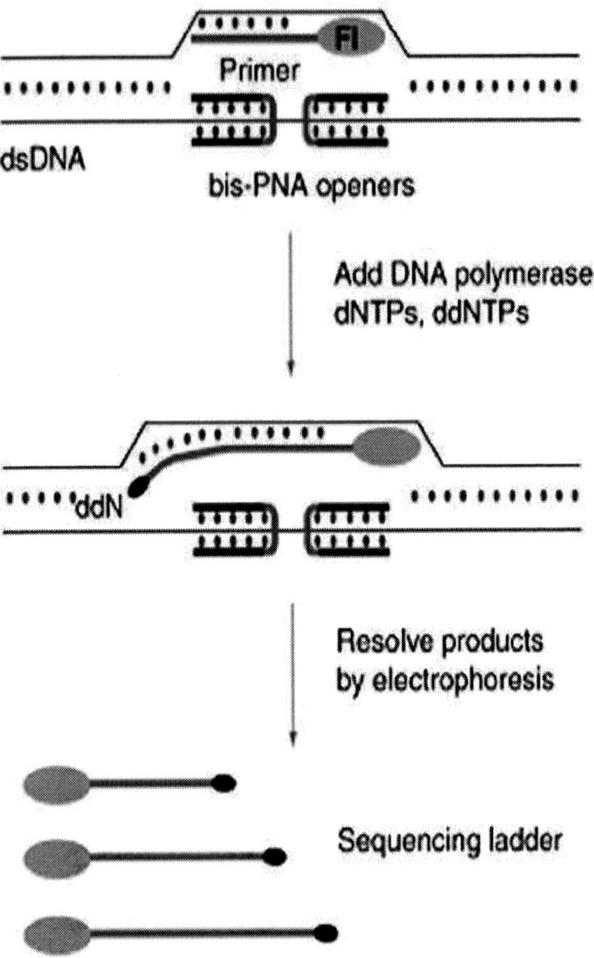


**Fig. 7.5:** Schematics of non-denaturing dsDNA sequencing xplaining the design and functioning of artificial primosome. Starting at the PD-loop, the sequencing ladder can be isothermally generated directly on dsDNA by DNA polymerase capable of strand-displacement replication, which extends primer in the presence of dNTPs and ddNTPs (chain terminators). Primer carries fluorescein (Fl) for optical detection of sequencing ladder after gel electrophoretic separation.

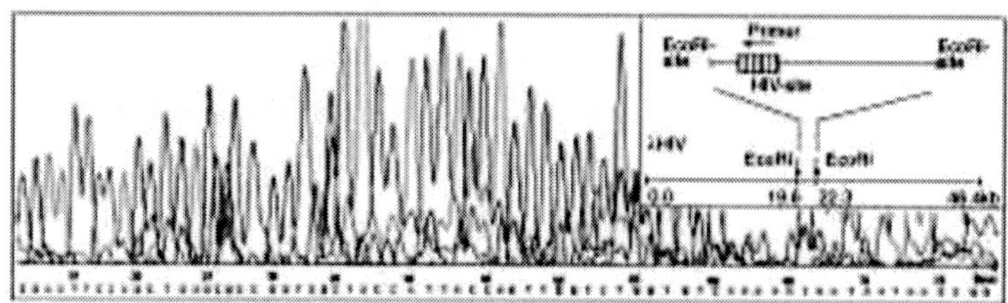

Fig. 7.6: Direct sequencing of 46.4 kb? HIV recombinant sDNA with Sequenase using an artificial primosome. Arrangement of the insertion with the HIV-1 PD-loop-forming site, AG2A2GCTACTG2AG2AGA (PNA binding sites in bold), within the ?gt11 vector is shown in the inset. A part of the sequence read thus obtained is presented (near 200 bp was sequenced in total; see refs. [34,35] for details). PNA openers: HLys2-TJTJ2T2J-(eg1)3-CT2C2TCT-LysNH2 and HLys2- TCTC2TC2-(eg1)3-J2TJ2TJT-LysNH2; oligonucleotide primer: fluorescein-5'GAG2A2GCTACTG2AG3'.

Therefore, the method promises to perform sequencing in the presence of unrelated DNA, thus evading target purification procedures. For instance, it may be employed for direct sequencing of inserts cloned within large vectors. Non-denaturing sDNA sequencing lets us avoid complications caused by folded loops that may form in denatured or ssDNA templates, but not in linear dsDNA templates. We anticipate the sequencing of this type could be used in various DNA diagnostics, e.g., in single nucleotide polymorphism (SNP) analyses.

The PD-loop-directed primer-extension reaction, when combined with a proper detection technique, allows the monitoring of subattomolar quantities of chosen DNA target on excessive DNA background. This approach has potential for ultrasensitive DNA assays.

## Hybridisation of Molecular Beacons to dsDNA

Molecular beacons are advantageous hybridisation probes carrying a fluorophore and a quencher at their termini thus enabling them to fluoresce upon hybridisation. To keep a quencher near a fluorophore in a free 'dark' state, DNA and PNA beacons were initially designed as hairpin-shaped molecules featuring a stem-and-loop structure. Later, it was realised that this structure is not obligatory for the functioning of these fluorescent probes and their stemless constructs have been developed. The workability of stemless DNA and PNA beacons can be attributed to the high flexibility of both sugar-phosphate and polyamide (pseudopeptide) backbones, and to strong hydrophobic interactions between quencher and fluorophore, which essentially act as a lock for closing the stemless constructs.

Molecular beacons have become very helpful tools for DNA diagnostics. However, to use them, DNA must be in a denatured single-stranded form to allow Watson-Crick pairing of molecular beacon to the target site. This requirement limits applications of molecular beacons. *Fig. 7.7* demonstrates that a dsDNA target, when locally exposed by PNA openers, may effectively hybridise with DNA and PNA beacons via the PD/PP-loop formation. Note here that PNA beacons are able to selectively detect target DNA in the presence of DNA-binding proteins. Conditions can be found where molecular beacons strongly discriminate the complementary versus mismatched dsDNA targets. Our findings open ways for applications of the molecular beacons methodology to non-denatured and/or non-deproteinized dsDNA analytes.

## Expert Opinion and Five-year View

With the presently widespread availability of numerous genomic sequences, DNA diagnostics may become an almost universal tool for biotechnological, clinical and forensic analyses. One of the major criteria, to which a candidate assay must satisfy, is high sequence specificity of detection. Examples given in this review demonstrate that the newly developed assays based on the PD-loop technology have such a potential.

Importantly, the selectivity of these assays is not compromised by excessive unrelated background. Moreover, they are able to successfully work with rather crude, non-purified samples.

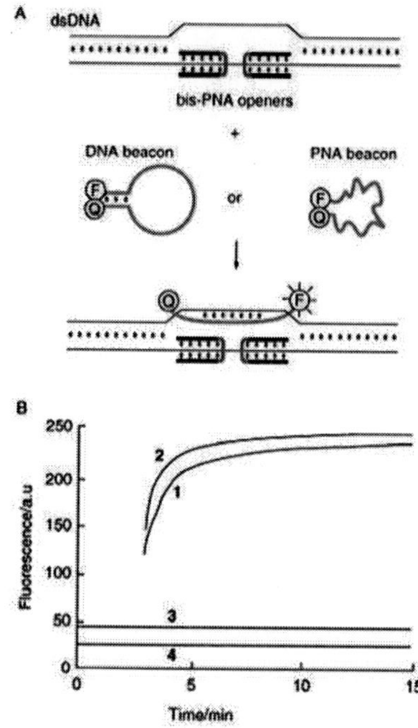


**Fig. 7.7:** Fluorescent detection of dsDNA target with molecular beacons. (A) Schematics of the hybridisation/detection assay: the formation of PDloop by DNA beacon or PP-loop by PNA beacon generates, upon illumination, a fluorescent response due to unfolding of molecular beacons in the hybridised state (Q: Quencher; F: Fluorophore). (B) Real-time kinetic responses of the item-forming DNA beacon (curve 1) and the stemless PNA beacon (curve 2) upon hybridisation to the complementary PD-loop-forming HIV-1 dsDNA site exposed by PNA openers (see legend to Fig. 6 for the sequences of DNA target and PNA openers). As a control, DNA and PNA beacons were added to target dsDNA not opened by PNA openers (lines 3 and 4, respectively). DNA beacon: Fl-GTACGTTggaagctactggagg TCGTAC3'- DABCYL (stem sequences are underlined; probe sequence is own in small letters; Fl: Fluorescein; DABCYL: Quencher); PNA beacon: Fl-Gluaagctactgga- Lys-Lys(DABCYL)-NH2 (note that the probe sequence of PNA beacon is shorter than that of DNA beacon, since PNA hybridises to ssDNA with very high affinity.

In the coming years, PD-loopbased assays will be used in various applications turning into a practical tool for DNA diagnostics. Besides the developments presented here, I expect several new applications of PD-loop technology to be elaborated. For example, it could be used for pre-gel hybridisation of a probe to duplex DNA, to design artificial nickase systems, to assemble other than circular topological labels and for site-directed DNA modifications. I also anticipate that PD-loops and similar constructs will find their place, as structural modules, in DNA nanotechnology and DNA biocomputing.

To fulfill these promises, two major improvements of PD-loop technology would be desirable. Note first that the current protocol of the PD-loop assembly requires purification of the locally opened dsDNA from PNA openers prior to probe hybridisation. While this shortcoming may limit certain applications of the PD-loop technology, it is not however unavoidable.

Indeed, the major reason for the removal of PNA openers is due to the sequence overlap between probes and openers, which is intrinsic to model systems we chose for our pilot studies. This overlap causes partial binding of probes with PNA openers and therefore, 'free' PNA openers should be removed from the PNADNA complexes formed at the stage of DNA opening to allow the proper hybridisation of the probe. It is reasonable to assume that choice of shorter probes and / or openers and use of other PD-loop-forming DNA sequences with longer gaps between the binding sites for PNA openers will allow simultaneous targeting of dsDNA with both openers and a probe, thus avoiding the need of such an additional purification step.

Although the shortest PNA openers used so far were (7+7)-mers, there are data demonstrating that pyrimidine bis-PNA as short as (5+5)-mer can stably invade a dsDNA target site.

Considering the superior binding affinity of PNA oligomers, shorter probes are certainly possible in those PD-loop applications when replacement of oligonucleotide with PNA probe is acceptable. For PNA probes, the optimal conditions for probe hybridisation are evidently compatible with low-salt concentrations required for PNA openers binding. In case of DNA or RNA probes, simultaneous use of both openers and a probe may require higher salt concentrations for optimal binding of the nucleic acid probe after some lag period needed for effective binding of PNA openers at low salt. However, high-salt conditions may cause the re-formation of DNA duplex between the opener's binding sites thus hindering the probe hybridisation, especially if these sites are separated by more than 10 bp. Use of bis-PNA openers appended with an extra PNA segment (could be called as tris-PNAs), which is complementary to a part of the spacer DNA sequence thus effectively preventing its closing, could help to resolve this problem. Therefore, though no serious problems are seen for the use of PD-loop technology in the 'no overlap' situations (our preliminary data prove this), such a possibility has to be experimentally demonstrated.

The second improvement necessary for further progress of PD-loop technology is elimination of sequence limitations.

The sequence limitations on the PD-loop-forming sites can be relaxed by the use of sequence-unrestricted pcPNAs as robust openers for the DNA double helix: pseudocomplementary pairs of this new PNA generation have recently demonstrated an ability to selectively target essentially any designated site on dsDNA by forming the PNA-DNA stranddisplacement complexes via the double-duplex invasion.

The PD-loop can be formed with the aid of a single pyrimidine bis-PNA opener and a pair of pc PNAs consisting of all four nucleobases. Then, only one short homopurine site, will be required for PD/PP-loop formation. Our preliminary data indicate that less sequence-restricted PP-loops of that kind can be formed by PNA probe with the aid of pyrimidine bis-PNA and pcPNAs openers. I am also convinced that the use of pcPNA openers in combination with pseudocomplementary oligonucleotides capable of invasion into mixed purine-pyrimidine dsDNA sequence at the edge of the duplex will finally result in the design of essentially sequence-universal PD-loops.

Besides these general improvements of PD-loop technology, other specific developments, which I anticipate to occur for each method described above, along with the new elaborations expected in this area. In closing, the 5- year view of the author is that the key advances of PD-loop technology during that period will result in the transition of this emerging DNA diagnostic methodology from infancy to maturity, ultimately transforming it into a practical biotechnological tool with many applications. The PD-loop technology, supported by the progress in the entire PNA field within the coming 5 years, has such a potential.

# 8

# Interactions of Multiple Genes

## Introduction

Organisms have thousands of genes, and in sexually reproducing organisms assortment of these genes are generally independent of each other. This means that the inheritance of an allele for yellow or green pea colour is unrelated to the inheritance of alleles for white or purple flowers. This phenomenon, known as "Mendel's second law" or the "Law of independent assortment", means that the alleles of different genes get shuffled between parents to form offspring with many different combinations.(Some genes do not assort independently, demonstrating genetic linkage, a topic discussed later in this article.)

Often different genes can interact in a way that influences the same trait. In the Blue-eyed Mary (Omphalodes verna), for example, there exists a gene with alleles that determine the colour of flowers: blue or magenta. Another gene, however, controls whether the flowers have colour at all: colour or white. When a plant has two copies of this white allele, its flowers are white— regardless of whether the first gene has blue or magenta alleles. This interaction between genes is called epistasis, with the second gene epistatic to the first.

Many traits are not discrete features (eg. purple or white flowers) but are instead continuous features (eg. human height and skin colour). These complex traits are the product of many genes. The influence of these genes is mediated, to varying degrees, by the environment an organism has experienced. The degree to which an organism's genes contribute to a complex trait is called heritability. Measurement of the heritability of a trait is relative — in a more variable environment, the environment has a bigger influence on the total variation of the trait. For example, human height is a complex trait with a heritability of 89 per cent in the United States. In Nigeria, however, where people experience a more variable access to good nutrition and health care, height has a heritability of only 62 per cent.

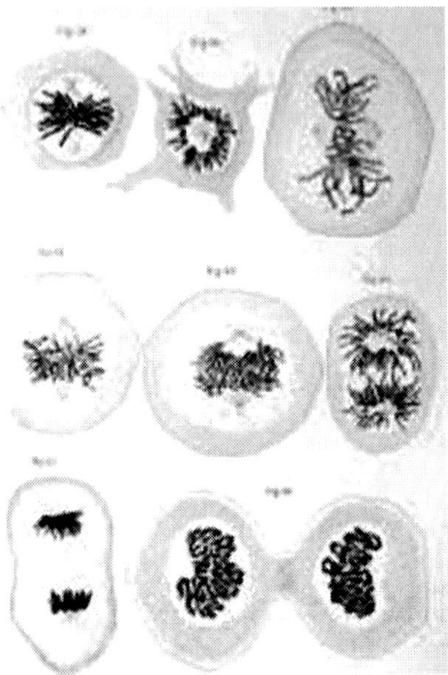

**Fig. 8.1:** Walther Flemming's 1882 diagram of eukaryotic cell division. Chromosomes are copied, condensed and organised. Then, as the cell divides, chromosome copies separate into the daughter cells.

## Molecular Basis for Inheritance

The molecular basis for genes is deoxyribonucleic acid (DNA). DNA is composed of a chain of nucleotides, of which there are four types: adenine (A), cytosine (C), guanine (G), and thymine (T). Genetic information exists in the sequence of these nucleotides, and genes exist as stretches of sequence along the DNA chain. Viruses are the only exception to this rule—sometimes viruses use the very similar molecule RNA instead of DNA as their genetic material.

DNA normally exists as a double-stranded molecule, coiled into the shape of a double-helix. Each nucleotide in DNA preferentially pairs with its partner nucleotide on the opposite strand: A pairs with T, and C pairs with G. Thus, in its two-stranded form, each strand effectively contains all necessary information, redundant with its partner strand. This structure of DNA is the physical basis for inheritance: DNA replication duplicates the genetic information by splitting the strands and using each strand as a template for synthesis of a new partner strand.

Genes are arranged linearly along long chains of DNA sequence, called chromosomes. In bacteria, each cell has a single circular chromosome, while eukaryotic organisms (which includes plants and animals) have their DNA arranged in multiple linear chromosomes. These DNA strands are often extremely long; the largest human chromosome, for example, is about 247 million base pairs in length. The DNA of a chromosome is associated with structural proteins that organise, compact, and control access to the DNA, forming a material called chromatin; in eukaryotes, chromatin is usually composed of nucleosomes, repeating units of DNA wound around a core of histone proteins. The full set of hereditary material in an organism (usually the combined DNA sequences of all chromosomes) is called the genome.

While haploid organisms have only one copy of each chromosome, most animals and many plants are diploid, containing two of each chromosome and thus two copies of every gene. The two alleles for a gene are located on identical loci of sister chromatids, each allele inherited from a different parent.

An exception exists in the sex chromosomes, specialised chromosomes many animals have evolved that play a role in determining the sex of an organism. In humans and other mammals, the Y chromosome has very few genes and triggers the development of male sexual characteristics, while the X chromosome is similar to the other chromosomes and contains many genes unrelated to sex determination. Females have two copies of the X chromosome, but males have one Y and only one X chromosome—this difference in X chromosome copy numbers leads to the unusual inheritance patterns of sex-linked disorders.

## Reproduction

When cells divide, their full genome is copied and each daughter cell inherits one copy. This process, called mitosis, is the simplest form of reproduction and is the basis for asexual reproduction. Asexual reproduction can also occur in multicellular organisms, producing offspring that inherit their genome from a single parent. Offspring that are genetically identical to their parents are called clones.

Eukaryotic organisms often use sexual reproduction to generate offspring that contain a mixture of genetic material inherited from two different parents. The process of sexual reproduction alternates between forms that contain single copies of the genome (haploid) and double copies (diploid). Haploid cells fuse and combine genetic material to create a diploid cell with paired chromosomes. Diploid organisms form haploids by dividing, without replicating their DNA, to create daughter cells that randomly inherit one of each pair of chromosomes. Most animals and many plants are diploid for most of their lifespan, with the haploid form reduced to single cell gametes.

Although they do not use the haploid/diploid method of sexual reproduction, bacteria have many methods of acquiring new genetic information. Some bacteria can undergo conjugation, transferring a small circular piece of DNA to another bacterium. Bacteria can also take up raw DNA fragments found in the environment and integrate them into their genome, a phenomenon known as transformation. This processes result in horizontal gene transfer, transmitting fragments of genetic information between organisms that would be otherwise unrelated.

## Recombination and Linkage

The diploid nature of chromosomes allows for genes on different chromosomes to assort independently during sexual reproduction, recombining to form new combinations of genes. Genes on the same chromosome would theoretically never recombine, however, were it not for the process of chromosomal crossover. During crossover, chromosomes exchange stretches of DNA, effectively shuffling the gene alleles between the chromosomes. This process of chromosomal crossover generally occurs during meiosis, a series of cell divisions that creates haploid germ cells that later combine with other germ cells to form child organisms.

The probability of chromosomal crossover occurring between two given points on the chromosome is related to the distance between them. For an arbitrarily long distance, the probability of crossover is high enough that the inheritance of the genes is effectively uncorrelated. For genes that are closer together, however, the lower probability of crossover means that the genes demonstrate genetic linkage-alleles for the two genes tend to be inherited together. The amounts of linkage between a series of genes can be combined to form a linear linkage map that roughly describes the arrangement of the genes along the chromosome.

## Gene Expression

Genes generally express their functional effect through the production of proteins, which are complex molecules responsible for most functions in the cell. Proteins are chains of amino acids, and the DNA sequence of a gene (through RNA intermediate) is used to produce a specific protein sequence. This process begins with the production of an RNA molecule with a sequence matching the gene's DNA sequence, a process called transcription.

This messenger RNA molecule is then used to produce a corresponding amino acid sequence through a process called translation. Each group of three nucleotides in the sequence, called a codon, corresponds to one of the twenty possible amino acids in protein—this correspondence is called the genetic code. The flow of information is unidirectional: information is transferred from nucleotide sequences into the amino acid sequence of proteins, but it never transfers from protein back into the sequence of DNA—a phenomenon Francis Crick called the central dogma of molecular biology.

The specific sequence of amino acids results in a unique three-dimensional structure for that protein, and the three-dimensional structures of protein are related to their function. Some are simple structural molecules, like the fibers formed by the protein collagen. Proteins can bind to other proteins and simple molecules, sometimes acting as enzymes by facilitating chemical reactions within the bound molecules (without changing the structure of the protein itself). Protein structure is dynamic; the protein hemoglobin bends into slightly different forms as it facilitates the capture, transport, and release of oxygen molecules within mammalian blood.

A single nucleotide difference within DNA can cause a single change in the amino acid sequence of a protein. Because protein structures are the result of their amino acid sequences, some changes can dramatically change the properties of a protein by destabilising the structure or changing the surface of the protein in a way that changes its interaction with other proteins and molecules. For example, sickle-cell anaemia is a human genetic disease that results from a single base difference within the coding region for the-globin section of haemoglobin, causing a single amino acid change that changes haemoglobin's physical properties. Sickle-cell versions of hemoglobin stick to themselves, stacking to form fibers that distort the shape of red blood cells carrying the protein. These sickle-shaped cells no longer flow smoothly through blood vessels, having a tendency to clog or degrade, causing the medical problems associated with this disease.

Some genes are transcribed into RNA but are not translated into protein products—these are called non-coding RNA molecules. In some cases, these products fold into structures which are involved in critical cell functions (eg. ribosomal RNA and transfer RNA). RNA can also have regulatory effect through hybridisation interactions with other RNA molecules (eg. microRNA).

## Nature Versus Nurture

Although genes contain all the information an organism uses to function, the environment plays an important role in determining the ultimate phenotype—a dichotomy often referred to as "nature vs. nurture." The phenotype of an organism depends on the interaction of genetics with the environment. One example of this is the case of temperature-sensitive mutations. Often, a single amino acid change within the sequence of a protein does not change its behaviour and interactions with other molecules, but it does destabilise the structure. In a high temperature environment, where molecules are moving more quickly and hitting each other, this results in the protein losing its structure and failing to function. In a low temperature environment, however, the protein's structure is stable and functions normally. This type of mutation is visible in the coat colouration of Siamese cats, where a mutation in an enzyme responsible for pigment production causes it to destabilise and lose function at high temperatures. The protein remains functional in areas of skin that are colder—legs, ears, tail, and face—and so the cat has dark fur at its extremities.

Environment also plays a dramatic role in effects of the human genetic disease phenylketonuria. The mutation that causes phenylketonuria disrupts the ability of the body to break down the amino acid phenylalanine, causing a toxic build-up of an intermediate molecule that, in turn, causes severe symptoms of progressive mental retardation and seizures. If someone with the phenylketonuria mutation follows a strict diet that avoids this amino acid, however, they remain normal and healthy.

## Gene Regulation

The genome of a given organism contains thousands of genes, but not all these genes need to be active at any given moment. A gene is expressed when it is being transcribed into mRNA (and translated into protein), and there exist many cellular methods of controlling the expression of genes such that proteins are produced only when needed by the cell. Transcription factors are regulatory proteins that bind to the start of genes, either promoting or inhibiting the transcription of the gene. Within the genome of Escherichia coli bacteria, for example, there exists a series of genes necessary for the synthesis of the amino acid tryptophan. However, when tryptophan is already available to the cell, these genes for tryptophan synthesis are no longer needed. The presence of tryptophan directly affects the activity of the genes—tryptophan molecules bind to the tryptophan repressor (a transcription factor), changing the repressor's structure such that the repressor binds to the genes.

The tryptophan repressor blocks the transcription and expression of the genes, thereby creating negative feedback regulation of the tryptophan synthesis process.

Differences in gene expression are especially clear within multicellular organisms, where cells all contain the same genome but have very different structures and behaviours due to the expression of different sets of genes. All the cells in a multicellular organism derive from a single cell, differentiating into variant cell types in response to external and intercellular signals and gradually establishing different patterns of gene expression to create different behaviours. As no single gene is responsible for the development of structures within multicellular organisms, these patterns arise from the complex interactions between many cells.

Within eukaryotes there exist structural features of chromatin that influence the transcription of genes, often in the form of modifications to DNA and chromatin that are stably inherited by daughter cells. These features are called "epigenetic" because they exist "on top" of the DNA sequence and retain inheritance from one cell generation to the next. Because of epigenetic features, different cell types grown within the same medium can retain very different properties. Although epigenetic features are generally dynamic over the course of development, some, like the phenomenon of paramutation, have multigenerational inheritance and exist as rare exceptions to the general rule of DNA as the basis for inheritance.

## Genetic Change

During the process of DNA replication, errors occasionally occur in the polymerisation of the second strand. These errors, called mutations, can have an impact on the phenotype of an organism, especially if they occur within the protein coding sequence of a gene. Error rates are usually very low—1 error in every 10–100 million bases—due to the "proofreading" ability of DNA polymerases. (Without proofreading error rates are a thousand-fold higher; because many viruses rely on DNA and RNA polymerases that lack proofreading ability, they experience higher mutation rates.) Processes that increase the rate of changes in DNA are called mutagenic: mutagenic chemicals promote errors in DNA replication, often by interfering with the structure of base-pairing, while UV radiation induces mutations by causing damage to the DNA structure. Chemical damage to DNA occurs naturally as well, and cells use DNA repair mechanisms to repair mismatches and breaks in DNA—nevertheless, the repair sometimes fails to return the DNA to its original sequence.

In organisms that use chromosomal crossover to exchange DNA and recombine genes, errors in alignment during meiosis can also cause mutations. Errors in crossover are especially likely when similar sequences cause partner chromosomes to adopt a mistaken alignment; this makes some regions in genomes more prone to mutating in this way. These errors create large structural changes in DNA sequence—duplications, inversions or deletions of entire regions, or the accidental exchanging of whole parts between different chromosomes (called translocation).

## Natural Selection and Evolution

Mutations produce organisms with different genotypes, and those differences can result in different phenotypes. Many mutations have little effect on an organism's phenotype, health, and reproductive fitness. Mutations that do have an effect are often deleterious, but occasionally mutations are beneficial. Studies in the fly Drosophila melanogaster suggest that if a mutation changes a protein produced by a gene, this will probably be harmful, with about 70 per cent of these mutations having damaging effects, and the remainder being either neutral or weakly beneficial.

Population genetics research studies the distributions of these genetic differences within populations and how the distributions change over time. Changes in the frequency of an allele in a population can be influenced by natural selection, where a given allele's higher rate of survival and reproduction causes it to become more frequent in the population over time. Genetic drift can also occur, where chance events lead to random changes in allele frequency.

Over many generations, the genomes of organisms can change, resulting in the phenomenon of evolution. Mutations and the selection for beneficial mutations can cause a species to evolve into forms that better survive their environment, a process called adaptation. New species are formed through the process of speciation, a process often caused by geographical separations that allow different populations to genetically diverge. The application of genetic principles to the study of population biology and evolution is referred to as the modern synthesis.

As sequences diverge and change during the process of evolution, these differences between sequences can be used as a molecular clock to calculate the evolutionary distance between them. Genetic comparisons are generally considered the most accurate method of characterising the relatedness between species, an improvement over the sometimes deceptive comparison of phenotypic characteristics. The evolutionary distances between species can be combined to form evolutionary trees—these trees represent the common descent and divergence of species over time, although they cannot represent the transfer of genetic material between unrelated species (known as horizontal gene transfer and most common in bacteria).

## Model Organisms and Genetics

Although geneticists originally studied inheritance in a wide range of organisms, researchers began to specialise in studying the genetics of a particular subset of organisms. The fact that significant research already existed for a given organism would encourage new researchers to choose it for further study, and so eventually a few model organisms became the basis for most genetics research. Common research topics in model organism genetics include the study of gene regulation and the involvement of genes in development and cancer.

Organisms were chosen, in part, for convenience—short generation times and easy genetic manipulation made some organisms popular genetics research tools. Widely used model organisms include the gut bacterium Escherichia coli, the plant Arabidopsis thaliana, baker's yeast (Saccharomyces cerevisiae), the nematode Caenorhabditis elegans, the common fruit fly (Drosophila melanogaster), and the common house mouse (Mus musculus).

## Medical Genetics Research

Medical genetics seeks to understand how genetic variation relates to human health and disease. When searching for an unknown gene that may be involved in a disease, researchers commonly use genetic linkage and genetic pedigree charts to find the location on the genome associated with the disease. At the population level, researchers take advantage of Mendelian randomisation to look for locations in the genome that are associated with diseases, a technique especially useful for multigenic traits not clearly defined by a single gene. Once a candidate gene is found, further research is often done on the same gene (called an orthologous gene) in model organisms. In addition to studying genetic diseases, the increased availability of genotyping techniques has led to the field of pharmacogenetics—studying how genotype can affect drug responses.

Although it is not an inherited disease, cancer is also considered a genetic disease. The process of cancer development in the body is a combination of events. Mutations occasionally occur within cells in the body as they divide. While these mutations will not be inherited by any offspring, they can affect the behaviour of cells, sometimes causing them to grow and divide more frequently. There are biological mechanisms that attempt to stop this process; signals are given to inappropriately dividing cells that should trigger cell death, but sometimes additional mutations occur that cause cells to ignore these messages. An internal process of natural selection occurs within the body and eventually mutations accumulate within cells to promote their own growth, creating a cancerous tumour that grows and invades various tissues of the body.

## Research Techniques

DNA can be manipulated in the laboratory. Restriction enzymes are a commonly used enzyme that cuts DNA at specific sequences, producing predictable fragments of DNA. The use of ligation enzymes allows these fragments to be reconnected, and by ligating fragments of DNA together from different sources, researchers can create recombinant DNA. Often associated with genetically modified organisms, recombinant DNA is commonly used in the context of plasmids—short circular DNA fragments with a few genes on them. By inserting plasmids into bacteria and growing those bacteria on plates of agar (to isolate clones of bacteria cells), researchers can clonally amplify the inserted fragment of DNA (a process known as molecular cloning). (Cloning can also refer to the creation of clonal organisms, through various techniques.)

DNA can also be amplified using a procedure called the polymerase chain reaction (PCR). By using specific short sequences of DNA, PCR can isolate and exponentially amplify a targeted region of DNA. Because it can amplify from extremely small amounts of DNA, PCR is also often used to detect the presence of specific DNA sequences.

## DNA Sequencing and Genomics

One of the most fundamental technologies developed to study genetics, DNA sequencing allows researchers to determine the sequence of nucleotides in DNA fragments. Developed in 1977 by Frederick Sanger and co-workers, chain-termination sequencing is now routinely used to sequence DNA fragments. With this technology, researchers have been able to study the molecular sequences associated with many human diseases.

As sequencing has become less expensive and with the aid of computational tools, researchers have sequenced the genomes of many organisms by stitching together the sequences of many different fragments (a process called genome assembly). These technologies were used to sequence the human genome, leading to the completion of the Human Genome Project in 2003. New high-throughput sequencing technologies are dramatically lowering the cost of DNA sequencing, with many researchers hoping to bring the cost of resequencing a human genome down to a thousand dollars.

The large amount of sequences available has created the field of genomics, research that uses computational tools to search for and analyse patterns in the full genomes of organisms. Genomics can also be considered a subfield of bioinformatics, which uses computational approaches to analyze large sets of biological data.

# 9

# Chromosome Structure and Morphology

## Introduction

Chromosomes are microscopic units containing organised genetic information, located in the nuclei of diploid and haploid cells (e.g. human somatic and sex cells), and are also present in one-cell non-nucleated organisms (unicellular microorganisms), like bacteria, which do not have an organised nucleus. The sum-total of genetic information contained in different chromosomes of a given individual or species are generically referred to as the genome.

In humans, chromosomes are structurally made of roughly equal amounts of proteins and DNA. Each chromosome contains a double-strand DNA molecule, arranged as a double helix, and tightly coiled and neatly packed by a family of proteins called histones. DNA strands are comprised of linked nucleotides. Each nucleotide has a sugar (deoxyribose), a nitrogenous base, plus one to three phosphate groups. Each nucleotide is linked to adjacent nucleotides in the same DNA strand by phosphodiester bonds. Phosphodiester is another sugar, made of sugar-phosphate. Nucleotides of one DNA strand link to their complementary nucleotide on the opposite DNA strand by hydrogen bonds, thus forming a pair of nucleotides, known as a base pair, or nucleotide base. Genes contain up to thousands of sequences of these base pairs. What distinguishes one gene from another is the sequence of nucleotides that code for the synthesis of a specific protein or portion of a protein. Some proteins are necessary for the structure of cells and tissues. Others, like enzymes, a class of active (catalyst) proteins, promote essential biochemical reactions, such as digestion, energy generation for cellular activity, or metabolism of toxic compounds. Some genes produce several slightly different versions of a given protein through a process of alternate transcription of bases pairs segments known as codons.

Amounts of autosomal chromosomes differ in cells of different species; but are usually the same in every cell of a given species. Sex determination cells (mature ovum and sperm) are an exception, where the number of chromosomes is halved. Chromosomes also differ in size. For instance, the smallest human chromosome, the sex chromosome Y, contains 50 million base pairs (bp), whereas the largest one, chromosome 1, contains 250 million base pairs. All 3 billion base pairs in the human genome are stored in 48 chromosomes. Human genetic information is therefore stored in 24 pairs of chromosomes (totaling 48), 24 inherited from the mother, and 24 from the father. Two of these chromosomes are sex chromosomes (chromosomes X and Y). The remaining 46 are autosomes, meaning that they are not sex chromosomes and are present in all somatic cells (i.e., any other body cell that is not a germinal cell for spermatozoa in males or an ovum in females). Sex chromosomes specify the offspring gender: normal females have two X-chromosomes and normal males have one X- and one Y-chromosome.

Each set of 24 chromosomes constitutes one allele, containing gene copies inherited from one of the progenitors. The other allele is complementary or homologous, meaning that they contain copies of the same genes and on the same positions, but originated from the other progenitor. As an example, every normal child inherits one set of copies of gene BRCA1, located on chromosome 13, from the mother and another set of BRCA1 from the father, located on the other allelic chromosome 13. Allele is a Greek-derived word that means "one of a pair," or any one of a series of genes having the same locus (position) on homologous chromosomes.

The first chromosome observations were made under light microscopes, revealing rod-shaped structures, in varied sizes and conformations commonly J-, or V-shaped in eukaryotic cells and ring-shaped chromosome in bacteria. Staining reveals a pattern of light and dark bands. Today those bands are known to correspond to regional variations in the amounts of the two nucleotide base pairs: Adenine-Thymine (A-T or T-A) in contrast with amounts of Guanine-Cytosine (G-C or C-G).

Genetic abnormalities and diseases occur when one of the following events happens: a) One chromosome copy is missing, b) Extra copies of a chromosome are present, c) A chromosome breaks and its fragment is fused into another chromosome (insertion), d) A fragment is deleted, e) A gene is transferred from one chromosome to another (translocation), f) Duplication of a chromosomal segment occurs, g) Inversion of a chromosomal segment occurs. Down syndrome, for instance, is caused by the presence of a third copy of chromosome 21.

In non-dividing cells, it is not possible to distinguish morphological details of individual chromosomes, because they remain elongated and entangled to each other. However, when a cell is dividing, i.e., undergoing mitosis, chromosomes become highly condensed and each individual chromosome occupies a well-defined spatial location.

Mitotic chromosomes present a constricted region, to which the spindle fibers attach during cellular division. Such constricted region, known as centromere or primary constriction, may be located in three different positions in chromosomes. Centromeric position allows the classification of chromosomes in three groups: (a) acrocentric: centromere lies very near one end; (b) metacentric: centromere at the middle, dividing the chromosome in two equal parts or arms; and (c) submetacentric: centromere near middle, but dividing chromosome in two unequal arms.

When a chromosome loses its centromere, it is known as acentric. As the centromere is essential for both division and retention of chromosome copies in the new cells, acentric chromosomes will not pass to the daughter cells during the parental cell division. Therefore, daughter cells will miss one chromosome in their karyotype. A karyotype map shows mitotic chromosomes in the mitotic phase, known as metaphase. In metaphase, chromosomes align in pairs. In a normal human karyotype, there are 22 pairs of autosomal chromosomes and two sex chromosomes (X and Y). Each pair of autosomal chromosomes contains two complementary or homologous chromosomes, a maternal and a paternal one.

Some chromosomes also present a secondary constriction that always appears at the same site. They are also useful, along with centromere position and chromosome size, for identifying and characterising individual chromosomes, in a karyotype.

Karyotype analysis was the first genetic screening utilised by geneticists to assess inherited abnormalities, like additional copies of a chromosome or a missing copy, as well as DNA content and gender of the individual. With the development of new molecular screening techniques and the growing number of identified individual genes, detection of other more subtle chromosomal mutations is now possible (e.g., determinations of gene mutations, levels of gene expression, etc). Such data allow scientists to better understand disease causation and to develop new therapies and medicines for those diseases.

## Mitotic Chromosomes

Isolated newt (Notophthalmus viridescens) chromosomes were studied by using micromechanical force measurement during nuclease digestion. Micrococcal nuclease and short-recognition-sequence blunt-cutting restriction enzymes first remove the native elastic response of, and then to go on to completely disintegrate, single metaphase newt chromosomes. These experiments rule out the possibility that the mitotic chromosome is based on a mechanically contiguous internal non-DNA (e.g., protein) "scaffold"; instead, the mechanical integrity of the metaphase chromosome is due to chromatin itself. Blunt-cutting restriction enzymes with longer recognition sequences only partially disassemble mitotic chromosomes and indicate that chromatin in metaphase chromosomes is constrained by isolated chromatin-crosslinking elements spaced by ?15 kb.

The question of how the millimeter-long chromatin fibers in eukaryote chromosomes are folded up into compacted metaphase chromatid remains open (1, 2). It has been proposed that metaphase chromatin is organised into loops attached to an underlying protein-rich structure called the metaphase scaffold (3–5). This scaffold has been visualised in the electron microscope by histone-depleting chromosomes; the resulting micrographs show? 40-kb DNA loops dangling from a protein-rich aggregate, which is the shape of the original chromosome (3). Isolated scaffolds were shown to contain nonhistone proteins (5) including topoisomerase II (6) and structural maintenance of chromosomes (7). Further studies indicated that the metaphase scaffold is helically folded (8) and contains AT-rich DNA sequences (9, 10). The basic "scaffold-loop" model suggested by these studies is widely accepted (11–14).

A fundamental question is whether the non-histone proteins of the scaffold are connected. Early studies stated that the scaffold was responsible for the basic shape of metaphase chromosomes and was essentially a fibrous network of nonhistone proteins (3), which could be isolated as a structurally independent stable entity (5). However, later discussions of this issue suggest that the question of whether the native scaffold is stabilised through protein–protein interactions is unresolved (6, 15). Furthermore, an old literature of whole-chromosome digestion experiments (16) appears inconsistent with a contiguous protein scaffold.

The alternative is that the proteins of the scaffold are not contiguously connected through the chromosome, and that DNA itself links the scaffold together. Here we report micromanipulation experiments that establish the non-DNA content of the mitotic chromosome to be mechanically disconnected, and therefore that mitotic chromosomes are best thought of as a chromatin network. Our experimental approach uses chromosome elastic response to report on structure and structural changes. We have recently developed techniques to carry out micromechanical experiments on isolated individual mitotic chromosomes (17, 18) using methods similar to those used in Nicklas' classic study of meiotic metaphase chromosome physical properties (19).

Observations that mitotic chromosomes can be stretched to five times their native length without permanent elongation

(17–19) suggested to us that the folded chromatin must be organised into a crosslinked chromatin network. To directly test this hypothesis, we carried out mechanical measurements on chromosomes during in situ digestion using DNA-cutting nuclease and restriction nucleases. Our aim was to monitor how chromosome elasticity and connectivity change dynamically as DNA cuts are made.

Our result is that, as DNA cuts are made within a mitotic chromosome by either micrococcal nuclease (MN) or frequently cutting restriction enzymes (REs), first chromosome elasticity is lost, and then the chromosome is completely dissolved. REs with higher specificity only partially reduce chromosome elasticity, showing that DNA and therefore chromatin are the mechanically contiguous components of mitotic chromosome structure and ruling out the possibility that the chromatin is organised by being tethered to a mechanically contiguous internal protein scaffold. Instead, we conclude that a mitotic chromosome is essentially a "network" of chromatin, with crosslinks about every 15 kb.

## Chromosome Micromanipulation

Micromechanical experiments were done with an inverted microscope (IX-70, 60X, 1.4 NA, Olympus, New Hyde Park, NY) equipped with two micromanipulators (MP-285, Sutter Instruments, Novato, CA) and a charge-coupled device camera controlled by a computer with labview and imaq (National Instruments, Austin, TX). Newt cells were grown in laboratory-made dishes designed to allow micropipettes to extract mitotic chromosomes.

After extraction, chromosomes were suspended between two micropipettes, one of which was used as a force transducer. This force-measuring micropipette was pulled and cut to have a bending force constant of ?0.1 nN/?m (1 nN = 10-9 N) and then mounted directly to the microscope stage to minimize mechanical noise. Force-extension experiments were done by moving the stiffer pipette at rates of ?0.01 ?m/sec, while observing the bending of the force-measuring pipette. After each experiment, the force-measuring pipette was calibrated as described in ref.

Previous experiments have shown that forces in the nanonewton range generate reversible stretching of mitotic chromosomes. The newt mitotic spindle is capable of generating nanonewton forces, and chromatids are often observed to be appreciably stretched during mitosis, particularly during anaphase. Therefore, the nanonewton forces used are comparable to forces applied to chromosomes *in vivo.*

All experiments were performed in the extracellular buffer, ?50 ?m above the cover glass on which the cells were cultured. The conditions for our experiments are therefore those of the extracellular medium (pH 7.5, ?100 mM net univalent salt, mainly Na+ and K+). A number of results indicate there is no appreciable restructuring of chromatin during or shortly after chromosome extraction into extracellular buffer. First, elasticity of chromosomes in the extracellular medium is close to that measured *in vivo*. Second, we observe no morphological change of chromosomes during extraction from cytoplasm to extracellular buffer. Third, we observe no gradual changes of physical properties of mitotic chromosomes from ?20-sec to 6-h exposure to extracellular buffer (likely due to the similarity of our extracellular buffer to solution conditions widely used to study chromatin structure). We, therefore, consider the extracted chromosomes to have essentially the same internal chromatin structure at the beginning of our experiments as they had inside the cell.

## Combined Enzymatic–Micromechanical Measurements

Once a single mitotic chromosome was isolated and held between two pipettes, the native elastic response of a chromosome was measured by observation of deflection of the force-measuring pipette while the other pipette is moved. Then, one of two types of experiments was done with an enzyme that was microinjected from a third 3-?m-diameter micropipette ?10 ?m from the chromosome (*Fig. 9.1 a*). In the first type of experiment, some tension is applied to the chromosome before exposing it to a DNA-cutting enzyme. Then, the enzyme is sprayed onto the chromosome while monitoring the tension (Fig. 9.1). Digital phase-contrast images were acquired at 10 frames/sec before, during, and after each enzyme exposure, recording pipette positions (therefore force and extension) and chromosome morphology. In the second type of experiment, the chromosome is sprayed with an enzyme with no applied tension. Then the force response of the chromosome is measured in the absence of any enzyme (*Fig. 9.2 on next page*).

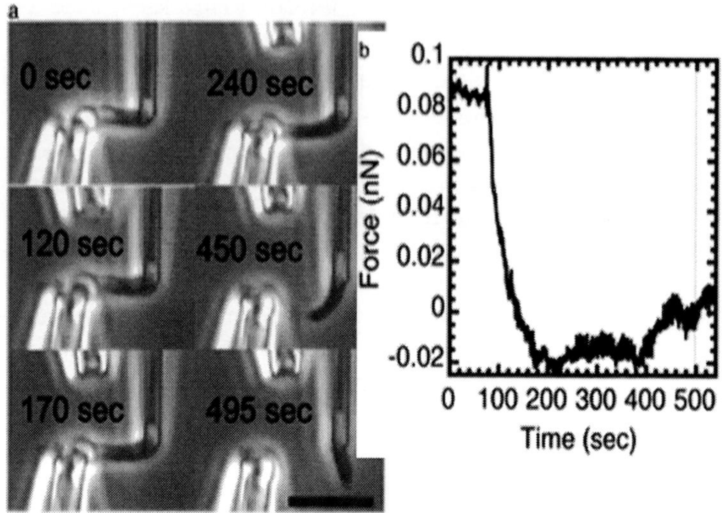


**Fig. 9.1:** Mechanical response of a newt mitotic chromosome microdigested with 1 nM MN in 60% PBS with 1 mM CaCl2. The chromosome was put under 0.1 nN of force before microdigestion. MN causes a relaxation of the initially applied force before any apparent change in chromosome morphology in phase contrast microscopy. Longer exposures lead to dissolution of the chromosome into unobservably small fragments, showing that a large-scale protein scaffold does not exist within mitotic chromosomes. (a) Phase images of the chromosome being digested by MN. The time in each image corresponds to the time axis of b. (Bar = 10 ?m.) A video is available at www.uic.edu/?jmarko/published/enz. (b) Time series of the force supported by the chromosome during the nuclease digestion. The thin vertical line indicates the time at which the chromosome was severed. Digestion was initiated at t = 60 sec.

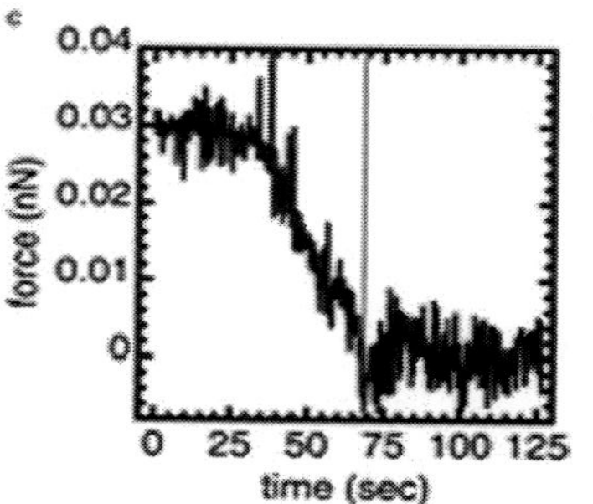


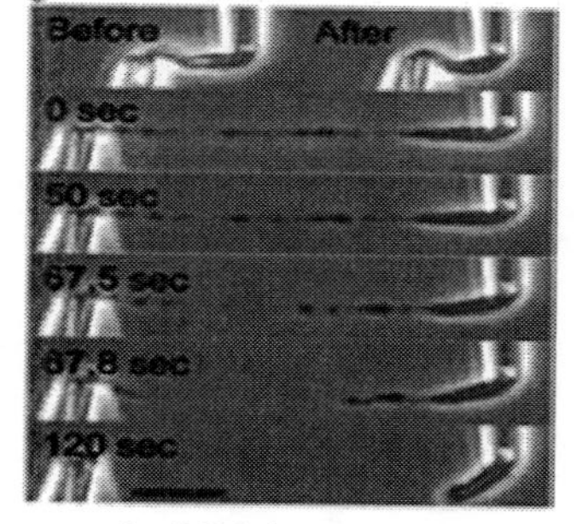


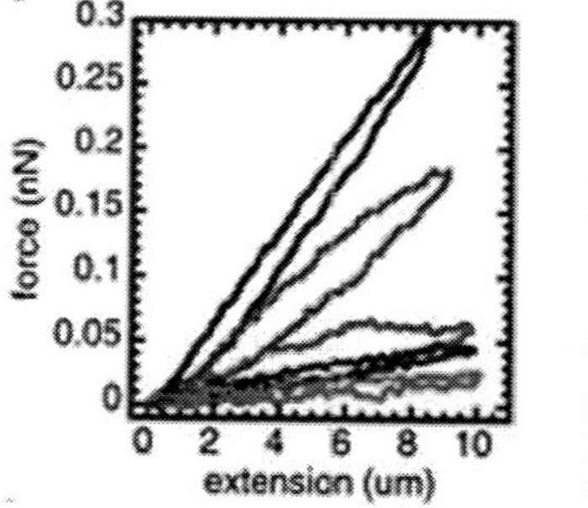


**Fig. 9.2** Partial digestion of a mitotic chromosome by MN shows it to behave as a chromatin network. A mitotic newt chromosome was digested with 10 nM MN for 90 sec without an applied tension. It was then subjected to four extension-retraction cycles in the absence of MN. The chromosome force constant was reduced after each extension–retraction cycle. After this, the mildly digested chromosome was extended to 40 ?m. The chromosome does not elongate homogeneously; instead, there are blobs connected by thin fibers. To test whether the thin fibers contain DNA, we exposed the extended blob-thin fiber structure to 10 nM MN while monitoring the force. The force relaxes in response to the exposure and the blob-thin fiber structure is cut through, indicating that the thin fibers contain DNA and that it is required to support the applied tension. (a) Force–extension response of a chromosome before (black) and after a 90-sec chromosome digestion with 10 nM MN. After digestion (without an applied tension), successive extension-relaxation cycles (blue, red, green, and purple) progressively reduce the force response; there is no longer reversible elasticity. (b) Top images show the chromosome unextended before (Left) and after (Right) 90-sec MN microdigestion; there is no obvious change in morphology under zero force. Lower images labelled with a time show the final cutting of the chromosome extended to 40 ?m, after the microdigestion and the extension-relaxation cycles of a. The t = 0 image shows the blob-link structure produced by microdigestion and stretching, whereas the t > 0 images show that the chromosome is completely cut by spraying with MN. (Bar = 10 ?m.) (c) Time series of the digestion experiment of b. The chromosome was extended by ?40 ?m and supported a force of ?30 pN. The microdigestion began at ?40 sec. The force relaxes to zero as the chromosome is cut. The low levels of force in this experiment are insufficient to break a single DNA molecule.

Spraying was controlled with a microinjection controller (World Precision Instruments, Sarasota, FL); all experiments used between 1- and 20-min sprays driven by 1,000-Pa pressure. When spraying is stopped, diffusion rapidly dissipates the sprayed reagent, abruptly stopping the reaction. In separate experiments with fluorescent dyes, we checked that the enzyme concentration at the chromosomes was within a factor of two of that inside the pipette (data not shown). Enzyme experiments were done with either small forces <1 nN or with zero force initially applied to the chromosome.

### DNA-Cutting Enzymes

MN and type II REs were used to induce cuts in double-stranded DNA. MN was prepared at 1–10 nM in 60 per cent PBS with 1 mM CaCl2. The REs used were AluI (AG?CT, Promega); HaeIII (GG?CC, Roche); Cac8I (GCN?NGC, New England Biolabs; N denotes "any base"); HincII [GT(T/C)? (A/G)AC, Promega; T/C denotes either T or C]; HindII [GT (T/C)?(A/G)AC, Roche Diagnostics, Indianapolis, IN], DraI (TTT?AAA, Promega), StuI (AGG?CCT, New England Biolabs); and PvuII (CAG?CTG, Promega), all of which produce blunt DNA cuts. REs were prepared at a concentration of 0.4–1.2 units/?l in appropriate reaction buffers (in each case, Tris?HCl, pH 7.5–8/50–100 mM NaCl/5–10 mM MgCl2). Because a relatively high RE concentration at room temperature in the chromosome experiments was used, each preparation of enzyme was tested by digestion of either 20 ng/?l pBR322 (Promega) or 10 ng/?l ?-DNA (Promega) at 25°C for 15, 30, and 60 min. Electrophoresis gel analysis showed that in each case, digestion was complete after 30 min, with no excess cutting or "star activity" (data not shown).

### MN Digestion of Mitotic Chromosomes

To investigate how MN affects chromosome elasticity and therefore structure, an extracted chromosome held by two micropipettes was elongated and retracted three times to determine its native elastic response. The chromosome was then extended so that that it supported ?0.1 nN. Before the spray of MN is initiated, the chromosome continuously supports this tension. After 60 sec (t = 60 sec; Fig. 9.1 b), the MN spray (1 nM) begins; the force supported by an elongated mitotic chromosome drops and is reduced to zero by MN after ?60 sec of spraying (t = 120 sec; Fig. 9.1 b), indicating reduction of elastic modulus to nearly zero. This softening occurs before any apparent morphological change (Fig. 9.1 a). After another 60 sec of exposure (t = 180 sec), the chromosome "thins," and after 500 sec, it is severed and subsequently completely dissolved. Similar results were found for four separate chromosomes with initial tensions ranging between 0.1 and 1 nN, and MN concentrations varied between 1 and 10 nM.

To further probe the structural change during the initial digestion by MN, a second type of experiment was done where an isolated chromosome with no applied tension was exposed to 10 nM MN for 90 sec. After this amount of digestion, no morphological change was observed. Then, the chromosome was repeatedly extended and retracted with no further MN spraying. Before digestion, a chromosome can be repeatedly extended and retracted without any change in its elastic response. By contrast, after this 90-sec MN digestion, repeated extension–retraction cycles are no longer reversible, with the force needed to double chromosome length dropping with each successive extension–retraction cycle (Fig. 9.2 a). Also, the chromosome no longer extends homogeneously: instead, relatively dense domains connected by thin fibers appear for >40-?m extensions (Fig. 9.2 b; t = 0 sec). To test whether the thin fibers contain DNA, they were extended and then sprayed with 10 nM MN. Immediately, the force relaxes and the thin fiber is cut (Fig. 9.2 b and c), indicating that its contiguous structural element is DNA and not a network of protein.

## RE Digestion of Mitotic Chromosomes

Experiments with blunt-cutting REs were done to estimate how often double-stranded cuts need to be made to disconnect the mitotic chromosome (REs cleave DNA at specific sequence and therefore with a given statistical frequency). Chromosomes were extended to ?1.5 times native length (force ?0.5 nN) and then were sprayed with a RE in appropriate reaction buffer. REs with either four- or six-base recognition sequences were used. The six-base REs varied in the number of six-base sequences recognised, which in turn varied the cut frequency.

Experiments with AluI produce results similar to MN, with force dropping to zero after 30 sec of spraying, and chromosome disintegration occurring after 200 sec of spraying (Fig. 9.3). HaeIII gives the same result (data not shown). These enzymes cut bare random-sequence DNA on average once every 44 = 256 bases or with a frequency 1/256 that of MN and still lead to complete disintegration of the chromosome. During AluI digestion, we do not observe any chromatin fragments.

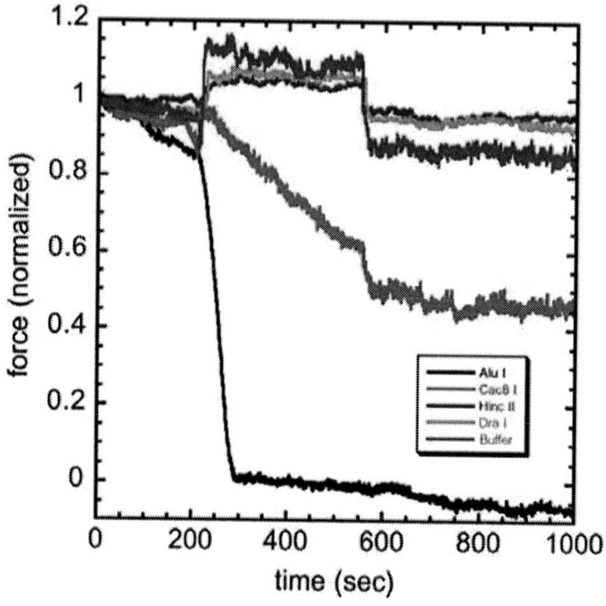


Fig. 9.3: REs with increasing specificity show decreasing effects on chromosome elastic response. Force data are before, during, and after 350-sec exposures to various REs; force is normalised to units of initial applied force, which ranged between 0.2 and 0.8 nN in the five separate experiments shown. AluI AG?CT (black) relaxes the force in ?30 sec; Cac8I GCN?NGC (red) only partially reduces the force. HincII GT(T/C)?(A/G)AC (blue) and DraI TTT?AAA (green) induce an increase in force during spraying, with a return to the original force when spraying stops (?600 sec), similar to spraying with reaction buffer and no enzyme (violet). These results indicate that chromatin-chromatin crosslinks occur roughly every 15 kb

Experiments with Cac8I were done to examine the effect of size of the recognition sequence length on enzyme activity. Cac8I's recognition sequence has the same statistical frequency on random-sequence DNA (1/44 = 1/256) as those of AluI and HaeIII but is spread over a six-base footprint. Cac8I exposure only partially reduces the applied force (Fig. 9.3). After spraying for 60 min, the force constant converges to 40 per cent of its native value (data not shown). The rate of force reduction by Cac8I is 1/10 that of AluI and HaeIII, indicating that the increase in recognition sequence size has reduced accessibility by ?10 times.

REs with high specificity do not reduce chromosome elasticity. HincII (Fig. 9.3) and HindII (data not shown), which cut random DNA once every 45 = 1,024 bases and one-fourth that of Cac8I, do not reduce chromosome elasticity. Experiments with DraI (TTT?AAA, Promega, Fig.9.3), StuI (AGG?CCT, New England Biolabs; data not shown), and PvuII (CAG?CTG, Promega; data not shown), which cut random DNA once every 46 = 4,096 bases, also do not produce observable force reduction.

The force increase observed during spraying with the less effective cutters ("step" response; Fig. 9.3) is due to the reaction buffers' divalent ions (6–10 mM Mg2+), which weakly and reversibly hypercondense mitotic chromosomes. Fig. 9.3 includes a force trace for reaction buffer with no enzyme, indicating that after spraying is complete, chromosome elastic response returns to its native value. The irreversible changes are therefore due to enzyme and not a buffer effect.

## Discussion

There Is No Contiguous Protein Scaffold Within Mitotic Chromosomes.

MN and 4-bp blunt REs eliminate a chromosome's ability to support a force. The reduction in elastic modulus occurs before any apparent morphological change occurs; long exposures "dissolve" the mitotic chromosome into unobservably small fragments. This result indicates that the mechanical integrity of the mitotic chromosome comes from DNA itself and shows there is no contiguous protein structure to which chromatin "loops" are merely tethered. Any protein "scaffold" should be made of relatively small isolated elements, connected together by chromatin (Fig. 9.4).

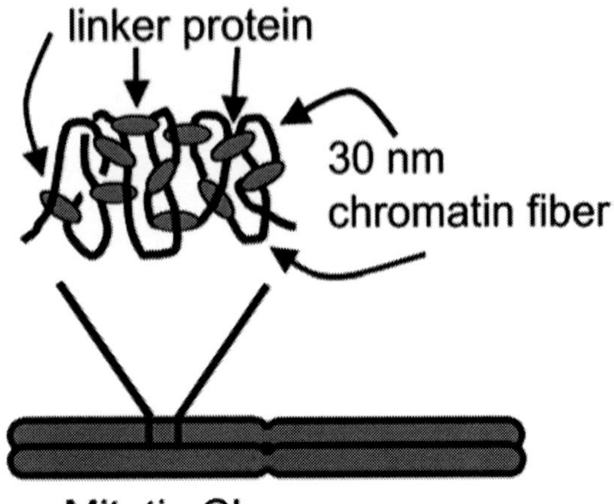


**Fig. 9.4:** Proposed "network" model of higher mitotic chromosome structure. The black lines represent 30-nm chromatin fiber, and the gray ovals represent proteins connecting the chromatin fiber to form a network-type structure. This model has a structure where the proteins crosslink chromatin, which maintains higher chromosome structure. When this structure is exposed to MN, cuts in the DNA (chromatin) between the crosslinks will be induced. The chromosome will no longer support an applied tension, which is what is observed experimentally. In addition, we estimate the average number of base pairs between crosslinks to be ?15 kb, based on the results from digesting the mitotic chromosome with various REs. Note that the crosslinks need not be homogeneously distributed through the chromatids.

This "network" picture is supported by experiments where short exposures to MN are made that do not disrupt chromosome morphology. Extension then leads the chromosome to break up into chromatin islands, attached by thin fibers (Fig. 9.2 b). These thin fibers can be cut by MN. This result, plus the greatly reduced elastic response (Fig. 9.2 a) and the weaker effects of more specific REs (Fig. 9.3), are all consistent with a network whose interconnects are randomly cut. In this proposed native state, chromosome elastic response is due to the self-adhesion and elastic response of the crosslinked folded chromatin [note that destabilizing chromatin–chromatin interactions reversibly decondenses whole chromosomes, combined with the elastic response of the protein crosslinks themselves.

One might imagine that part of the internal protein structure of the chromosome is lost as a result of extraction, either as a result of dissociation under dilute conditions or as a consequence of a lack of energy sources or other cofactors not present in the extracellular medium. However, such "lost" protein structures do not contribute significantly to the mechanical properties of mitotic chromosomes, because their bending and stretching properties in vivo are similar to those measured in the extracellular medium. Repetition of the experiments of this paper in mitotic cell extracts may be able to further address this question.

Chromatin in Mitotic Chromosomes Is Crosslinked Roughly Every 15 kb.

MN and REs with a 4-bp recognition sequence lead to compete disintegration of the chromosome. Four-base-pair REs cut bare random-sequence DNA on average once every 44 = 256 bases or with a frequency 1/256 that of MN. Because the access of REs to DNA is reduced in chromatin by ?10-fold, we conclude that roughly one cut every 2.5 kb (12 nucleosomes) is sufficient to completely disassemble a mitotic chromosome.

REs with higher specificities were used to reduce the number of DNA cuts made in a mitotic chromosome, using gradually rarer recognition sequences. A complication is that longer recognition sequences will be less accessible along the short internucleosomal linker DNAs in chromatin. To control for this effect, Cac8I was used. This enzyme recognises a 6-bp recognition sequence but has 16-fold degeneracy and therefore cuts with the same statistical frequency as AluI and HaeIII. The rate of force reduction by Cac8I is 1/10 that of AluI and HaeIII, indicating that the 2-bp increase in recognition sequence size reduces accessibility by ?10 times. We therefore estimate that Cac8I can cut metaphase chromatin about once every 25 kb. For a network architecture, reduction in force constant to 40 per cent of native requires cutting of 60 per cent of the links. On the basis of the estimate that Cac8I cuts once per 25 kb, we conclude that the length of chromatin between crosslinks in the mitotic chromosome is ?15 kb. We emphasise that this estimate is approximate and that it may be affected by a number of factors, including kinetics of RE access to DNA.

We, therefore, expect that six-base blunt-cutting enzymes (i.e., with the same DNA access as Cac8I) with lower cutting frequencies will produce proportionally smaller force relaxations. Given our force resolution of ?0.05 nN, no observable change is expected. This is the experimental result observed for all of the higher specificity enzymes we used: HincII (Fig. 9.3), HindII (data not shown), DraI (Fig. 9.3), StuI (data not shown), and PvuII (data not shown).

The large difference in activity of Cac8I and AluI is somewhat surprising, because the recognition sequences differ by only 2 bp in length, a small fraction of the 20-40 bp between nucleosomes. It is possible that shape and size of these enzymes play significant roles. Alternately, only subregions of internucleosomal linker DNA may be accessible between the footprints of other DNA-bound proteins such as topoisomerases and SMCs (structural maintenance of chromosome proteins).

Thus, sufficiently frequent DNA cuts in a mitotic chromosome completely disassemble it, proving that a contiguous protein scaffold does not exist within mitotic chromosomes. Instead, our results indicate that DNA itself provides the mechanical integrity of chromosomes, and that mitotic chromosomes can be considered to be crosslinked networks of chromatin (Fig. 9.4). Our result that chromatin is crosslinked every ?15 kb is comparable to the sizes of DNA fragments obtained from classical chromatin "loop size" analyses. Our experiments reveal that the classical loops are better thought of as network links, because in the mitotic state, the non-DNA portions of the loop "bases" are not solidly connected to one another. It should be noted that if the construct of Fig. 9.4 were treated so that the network nodes adhered to one another, it is natural to expect that apparent loops emanating from an apparent protein-rich condensate would be observed after histone removal, explaining the classical loop visualisation experiments. Finally, we emphasise that a network as in Fig. 9.4 can result from a hierarchical folding process. Further insight may come from analysis of the chromatin fragment size distributions produced by digestion by various REs.

## Ploidy: Diplod and Haploid Chromosome Number of Human Somatic (2n) and Sex Cells (n)

Genetics is rife with potentially confusing scientific terminology associated with chromosomes and chromosome number. What is ploidy, and which cells are haploid? Which are

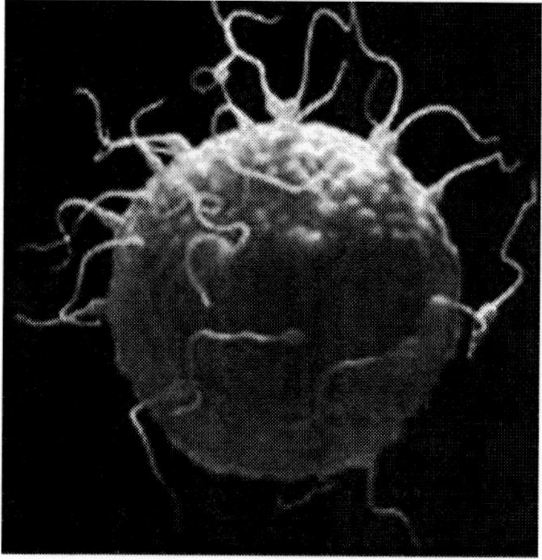

Fig. 9.5: Sperm and Egg Gametes: Haploid Sex Cells

diploid? What are homologous chromosomes? What are duplicated chromosomes? The intent of this article series is to clarify the lingo of genetics.

## Somatic Cells Diplod Number

Most of the cells in our bodies are somatic, or non sex cells, and have a diplod (2n) chromosome number. Diploid means that chromosomes come in pairs called homologues. Every somatic cell in your body has 46 chromosomes. You received a set of 23 from your mother's egg and a matching set of 23 via your father's sperm, and now these chromosomes are the genetic material inside nearly every cell of your body.

## Ploidy and Mitosis

Mitosis is cell division that results in the duplication of somatic cells; the 'daughter cells' genetic copies of the 'parent cell'. This cell multiplication allows for replacement of old cells, tissue repair, growth and development.

In order to divide and produce identical 'daughter cells', a 'parent cell' must first duplicate (replicate) its genetic material (DNA). So, prior to cell division, somatic diplod cells have duplicated chromosomes. As soon as the cell divides, each new cell has a complete diplod number of 46, unduplicated chromosomes.

## Haploid Number

Most, but not all, of the cells in our body and diploid. Sex cells, also called gametes (sperm or eggs), have half the number of chromosomes as do somatic cells. These gametes are referred to as being haploid (1n).

## Ploidy and Meiosis

In sexually reproducing organisms, gametes are produced by another method of cell division called meiosis. Meiosis is much more complex than mitosis. Whereas mitosis involves the duplication and subsequent division of chromosomes, meiosis involves two divisions of genetic material.

Gametes have half the number of chromosomes than the progenitor cell that they arose from. These haploid sex cells arise in specialised reproductive tissue called the gonads. Ovaries (female gonads) and testes (male gonads) are the sites of meiosis.

## Human Life Cycle: Haploid to Diploid

The merging of haploid sperm and egg at fertilisation brings the chromosome count back to 2n diploid number necessary for a zygote to have complete genetic information; 2 sets of genetic instructions in 23 pairs of chromosomes.

## Non-sex Genes 'Link to Gay Trait'

Multiple genes—and not just the sex chromosomes—are important in sexual orientation, say US scientists.

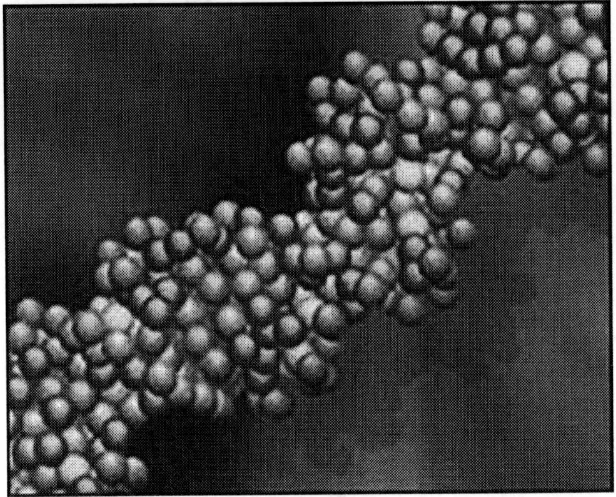

**Fig. 9.6**

A University of Illinois team, which has screened the entire human genome, say there is no one 'gay' gene.

Writing in the journal Human Genetics, they said environmental factors are also likely to be involved.

The findings add to the debate over whether sexual orientation is a matter of choice. Campaigners say equality is the more important issue.

## Non-sex Genes

Much of the past genetic research into male homosexuality had focussed solely on the X chromosome, passed down to boys by their mother, according to lead researcher Dr Brian Mustanski.

His team looked at all 22 pairs of non-sex chromosomes of 456 individuals from 146 families with two or more gay brothers.

Dr. Mustanski said the next step would be to see if the findings could be confirmed by further studies, and to identify the particular genes within the newly discovered sequences that are linked to sexual orientation.

"Our study helps to establish that genes play an important role in determining whether a man is gay or heterosexual," he said, but added that other factors were also important.

"Sexual orientation is a complex trait. There is no one 'gay' gene.

"Our best guess is that multiple genes, potentially interacting with environmental influences, explain differences in sexual orientation."

Alan Wardle from the gay rights charity Stonewall said: "It's an interesting study that contributes further to the debate.

"Regardless of whether sexual orientation is determined by nature or nurture or both, the most important thing is that lesbians and gay men are treated equally and are allowed to live their life without discrimination."

## Chromosome Aberrations

Chromosome aberrations are departures from the normal set of chromosomes either for an individual or from a species. They can refer to changes in the number of sets of chromosomes (ploidy), changes in the number of individual chromosomes (somy), or changes in appearance of individual chromosomes through mutation-induced rearrangements. They can be associated with genetic diseases or with species differences.

## Trisomy

Chromosome number, size, and shape (X-shaped or V-shaped) are fixed for each species. In most animals, chromosomes are present in pairs, called homologous pairs, carrying similar genes. Each chromosome pair carries a distinctive set of genes. Genes code for proteins, and the amount of protein produced in a cell from a particular gene is proportional to the number of functional gene copies present.

Trisomy refers to having three copies of one chromosome. It arises through the chromosomal accident of non-disjunction during meiosis, which sends two copies of a particular chromosome into a sperm or egg, rather than one. An individual with a trisomy who survives to be born produces more of the gene products encoded on the trisomic chromosome. The resulting genic imbalance almost always severely impairs growth. Trisomy of the twenty-first chromosome, the smallest in humans, is the cause of Down syndrome, which is associated with mental retardation, congenital heart disease, accelerated aging, and characteristic facial features. Trisomy that occurs after fertilisation, during fetal development, results in a "mosaic" individual with only some trisomic cells in the body. Such individuals may display some but not all of the features of the syndrome.

Trisomics for different chromosomes result in different abnormal characteristics. In humans, trisomies 13 and 18 are associated with different birth defects. While these individuals do not live long after birth, trisomies for any of the other non-sex chromosomes die before birth. Monosomies (having only one copy) for any chromosome also do not survive fetal existence, except for the sex chromosomes X and Y. Sex chromosome trisomies and monosomy for the X chromosome are associated with less severe effects on the phenotype.

## Translocations

Translocations are the result of a chromosomal-level mutation, with two different (non-homologous) chromosomes breaking and rejoining, placing the genes from one part of the one chromosome with part of the second chromosome, and vice versa. The number of genes is unchanged. Occasionally, the breakpoint mutation interrupts and inactivates the gene located at that chromosomal site. In other cases, the juxtaposition of new deoxyribonucleic acid (DNA) sequences from the other chromosome next to a gene at the breakpoint results in inappropriate expression. This action may activate an oncogene, for example. The "Philadelphia chromosome" is a translocation that fuses parts of chromosomes 9 and 22, which produces a new gene product that functions as an oncogene called Abl, which is implicated in chronic myelogenous leukaemia.

Inherited translocations are passed through generations in a codominant fashion. Since one copy of each chromosome remains normal, both parent and progeny with such a translocation are heterozygous, or "balanced" carriers. Half their gametes will include one copy of each gene, either on the translocated chromosomes or their normal homologs. The other half, however, are unbalanced with some combination of translocated and normal homologs. The result is that the gamete has two copies of some genes, but no copies of other genes, from the translocated chromosomes. Such an "unbalanced" gamete, if it takes part in fertilization, often disrupts development so greatly that the individual does not survive to be born. If the number of unbalanced genes is low, however, children may be born, but often they have growth defects and mental retardation. Couples with recurrent spontaneous abortions may have one partner carrying a balanced translocation. Thus, gene copy number determines the specific phenotypes associated with a translocation, or with any chromosome aberration. Extreme examples of the importance of gene number are triploidy (3n 69 for humans) and tetraploidy (4n 92). These individuals nearly always die early in fetal life and are detected only in the remains of early spontaneous abortions. However, intolerance of polypoidy may be a mammalian phenomenon. It is common among plants, and frogs that are triploid are both viable and fertile.

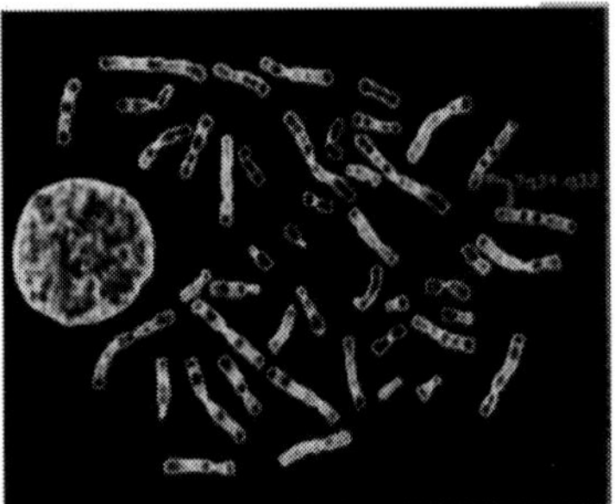

**Fig. 9.7**

Robertsonian translocations are a special class that result from the fusion of two V-shaped chromosomes at their centromere ends to form a single X-shaped chromosome. Individuals who are balanced for this translocation have forty-five chromosomes, but are otherwise normal. However, during gamete formation, some gametes will become unbalanced, and their progeny are at risk for being aneuploid (without the correct set of genes). If the two fused chromosomes are homologs, then the risk is 100 per cent that the zygote will be aneuploid since it will either have too few or too many genes. If nonhomologs are fused, the risk is usually 50 per cent. About 5 per cent of Down syndrome cases are caused by Robertsonian translocations.

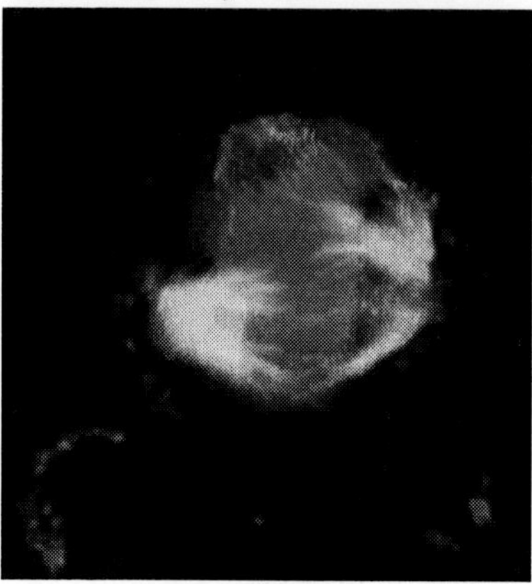

Fig. 9.8

Chromosome fusions or exchanges at the centromere position can often be correlated with differences between closely related species. The great apes (chimpanzees and gorillas) have forty-eight chromosomes, for example. Their chromosome constitution differs from humans only in having one fewer small X-shaped chromosome pair and two additional small V-shaped pairs. At some point in our past, the V-shaped pairs fused at the centromeres to form the X-shaped pairs. Using DNA sequences that are conserved among vertebrate species as position markers, it is possible to create comparative maps of conserved blocks of genes. All mammalian X chromosomes are alike in the genes present, for instance. Thus, changes visible at the chromosome level are useful markers to follow the evolutionary relatedness of different species. Further comparison of conserved sequences suggests the vertebrate genome is a tetraploid (four-copy) version of the invertebrate genome.

## Polyploidy

Changes in ploidy are of evolutionary importance in the flowering plants (angiosperms). Tetraploid plants often grow faster and larger than the diploid plants they derive from, and tend to be selected for agriculture. Alfalfa, coffee, wheat, peanuts, and potatoes are some examples. Commercially grown strawberries are octaploids (eight chromosome sets). Triploid plants, formed by crossing tetraploid with related diploid species, are almost always sterile because their aneuploid seeds abort. Seedless watermelons and bananas are examples of this technique used to improve fruit for human consumption.

An example of plant polyploidy that affected human civilisation and history is the origin of wheat. Modern bread wheat, cultivated for about eight thousand years, is a hexaploid with 2n 42, formed by sequential hybrids formed among three related grass species, each with 2n 14. To complete the circle, modern hybridisers have created a new species, Triticale, by crossing the ancestral Emmer wheat (2n 28) with rye (2n 14) and then doubling the chromosome number to 42 to take advantage of strong wheat growth with the high lysine content of rye. DNA analysis is scrutinising the hybrid origins of many cultivated plants to identify their ancestors.

# 10

# Gene Mutation

## Introduction

A gene mutation is a permanent change in the DNA sequence that makes up a gene. Mutations range in size from a single DNA building block (DNA base) to a large segment of a chromosome.

Gene mutations occur in two ways: they can be inherited from a parent or acquired during a person's lifetime. Mutations that are passed from parent to child are called hereditary mutations or germline mutations (because they are present in the egg and sperm cells, which are also called germ cells). This type of mutation is present throughout a person's life in virtually every cell in the body.

Mutations that occur only in an egg or sperm cell, or those that occur just after fertilisation, are called new (de novo) mutations. De novo mutations may explain genetic disorders in which an affected child has a mutation in every cell, but has no family history of the disorder.

Acquired (or somatic) mutations occur in the DNA of individual cells at some time during a person's life. These changes can be caused by environmental factors such as ultraviolet radiation from the sun, or can occur if a mistake is made as DNA copies itself during cell division. Acquired mutations in somatic cells (cells other than sperm and egg cells) cannot be passed on to the next generation.

Mutations may also occur in a single cell within an early embryo. As all the cells divide during growth and development, the individual will have some cells with the mutation and some cells without the genetic change. This situation is called mosaicism.

Some genetic changes are very rare; others are common in the population. Genetic changes that occur in more than 1 per cent of the population are called polymorphisms. They are common enough to be considered a normal variation in the DNA. Polymorphisms are responsible for many of the normal differences between people such as eye colour, hair colour, and blood type. Although many polymorphisms have no negative effects on a person's health, some of these variations may influence the risk of developing certain disorders.

## What is DNA?

DNA, or deoxyribonucleic acid, is the hereditary material in humans and almost all other organisms. Nearly every cell in a person's body has the same DNA. Most DNA is located in the cell nucleus (where it is called nuclear DNA), but a small amount of DNA can also be found in the mitochondria (where it is called mitochondrial DNA or mtDNA).

The information in DNA is stored as a code made up of four chemical bases: adenine (A), guanine (G), cytosine (C), and thymine (T). Human DNA consists of about 3 billion bases, and more than 99 per cent of those bases are the same in all people. The order, or sequence, of these bases determines the information available for building and maintaining an organism, similar to the way in which letters of the alphabet appear in a certain order to form words and sentences.

DNA bases pair up with each other, A with T and C with G, to form units called base pairs. Each base is also attached to a sugar molecule and a phosphate molecule. Together, a base, sugar, and phosphate are called a nucleotide. Nucleotides are arranged in two long strands that form a spiral called a double helix. The structure of the double helix is somewhat like a ladder, with the base pairs forming the ladder's rungs and the sugar and phosphate molecules forming the vertical sidepieces of the ladder.

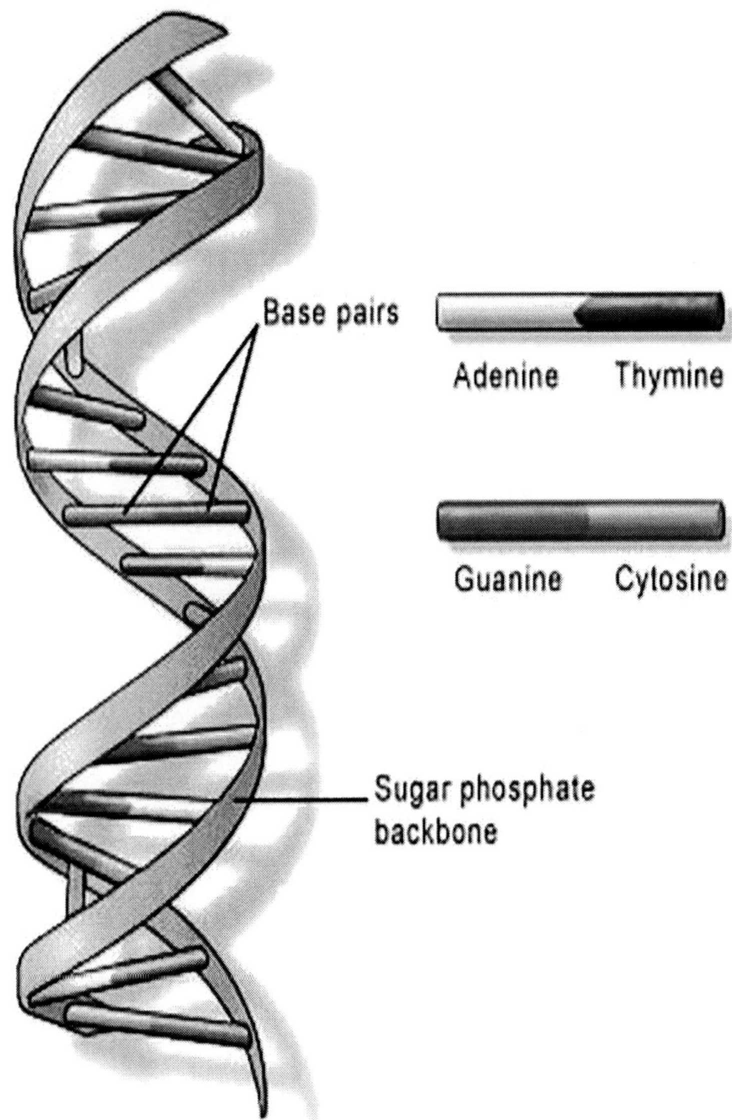


**Fig. 10.1**: DNA is a double helix formed by base pairs attached to a sugar-phosphate backbone

An important property of DNA is that it can replicate, or make copies of itself. Each strand of DNA in the double helix can serve as a pattern for duplicating the sequence of bases. This is critical when cells divide because each new cell needs to have an exact copy of the DNA present in the old cell.

## How Genes Work

### What are proteins and what do they do?

Proteins are large, complex molecules that play many critical roles in the body. They do most of the work in cells and are required for the structure, function, and regulation of the body's tissues and organs.

Proteins are made up of hundreds or thousands of smaller units called amino acids, which are attached to one another in long chains. There are 20 different types of amino acids that can be combined to make a protein. The sequence of amino acids determines each protein's unique 3-dimensional structure and its specific function.

### How do genes direct the production of proteins?

Most genes contain the information needed to make functional molecules called proteins. (A few genes produce other molecules that help the cell assemble proteins.) The journey from gene to protein is complex and tightly controlled within each cell. It consists of two major steps: transcription and translation. Together, transcription and translation are known as gene expression.

During the process of transcription, the information stored in a gene's DNA is transferred to a similar molecule called RNA (ribonucleic acid) in the cell nucleus. Both RNA and DNA are made up of a chain of nucleotide bases, but they have slightly different chemical properties. The type of RNA that contains the information for making a protein is called messenger RNA (mRNA) because it carries the information, or message, from the DNA out of the nucleus into the cytoplasm.

Translation, the second step in getting from a gene to a protein, takes place in the cytoplasm. The mRNA interacts with a specialised complex called a ribosome, which "reads" the sequence of mRNA bases. Each sequence of three bases, called a codon, usually codes for one particular amino acid. (Amino acids are the building blocks of proteins.) A type of RNA called transfer RNA (tRNA) assembles the protein, one amino acid at a time. Protein assembly continues until the ribosome encounters a "stop" codon (a sequence of three bases that does not code for an amino acid).

The flow of information from DNA to RNA to proteins is one of the fundamental principles of molecular biology. It is so important that it is sometimes called the "central dogma."

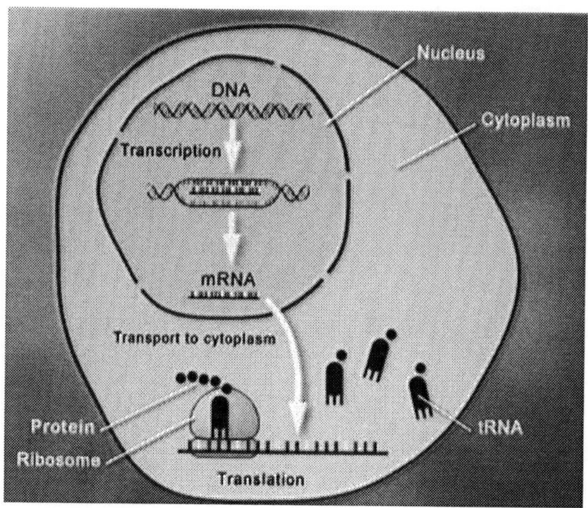


Fig. 10.2: Through the processes of transcription and translation, information from genes is used to make proteins

## Can genes be turned on and off in cells?

Each cell expresses, or turns on, only a fraction of its genes. The rest of the genes are repressed, or turned off. The process of turning genes on and off is known as gene regulation. Gene regulation is an important part of normal development. Genes are turned on and off in different patterns during development to make a brain cell look and act different from a liver cell or a muscle cell, for example. Gene regulation also allows cells to react quickly to changes in their environments. Although we know that the regulation of genes is critical for life, this complex process is not yet fully understood.

Gene regulation can occur at any point during gene expression, but most commonly occurs at the level of transcription (when the information in a gene's DNA is transferred to mRNA). Signals from the environment or from other cells activate proteins called transcription factors. These proteins bind to regulatory regions of a gene and increase or decrease the level of transcription. By controlling the level of transcription, this process can determine the amount of protein product that is made by a gene at any given time.

## How do cells divide?

There are two types of cell division: mitosis and meiosis. Most of the time when people refer to "cell division," they mean mitosis, the process of making new body cells. Meiosis is the type of cell division that creates egg and sperm cells.

Mitosis is a fundamental process for life. During mitosis, a cell duplicates all of its contents, including its chromosomes, and splits to form two identical daughter cells. Because this process is so critical, the steps of mitosis are carefully controlled by a number of genes. When mitosis is not regulated correctly, health problems such as cancer can result.

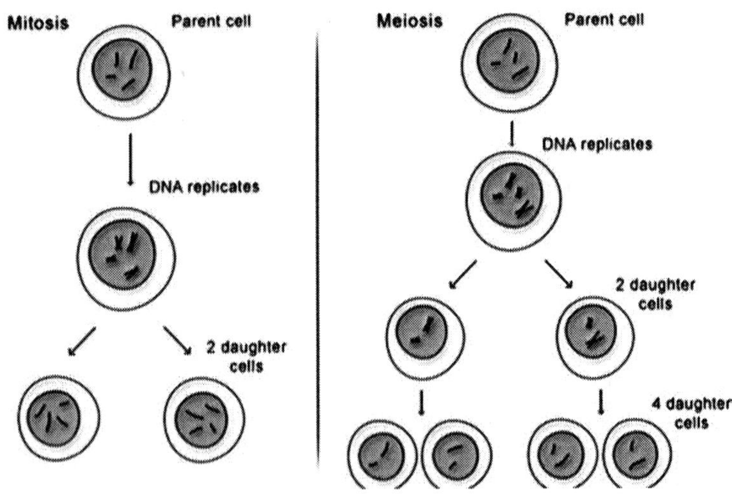


**Fig. 10.3:** Mitosis and meiosis, the two types of cell division

The other type of cell division, meiosis, ensures that humans have the same number of chromosomes in each generation. It is a two-step process that reduces the chromosome number by half—from 46 to 23—to form sperm and egg cells. When the sperm and egg cells unite at conception, each contributes 23 chromosomes so the resulting embryo will have the usual 46. Meiosis also allows genetic variation through a process of DNA shuffling while the cells are dividing.

## How do genes control the growth and division of cells?

A variety of genes are involved in the control of cell growth and division. The cell cycle is the cell's way of replicating itself in an organised, step-by-step fashion. Tight regulation of this process ensures that a dividing cell's DNA is copied properly, any errors in the DNA are repaired, and each daughter cell receives a full set of chromosomes. The cycle has checkpoints (also called restriction points), which allow certain genes to check for mistakes and halt the cycle for repairs if something goes wrong.

If a cell has an error in its DNA that cannot be repaired, it may undergo programmed cell death (apoptosis) (illustration). Apoptosis is a common process throughout life that helps the body get rid of cells it doesn't need. Cells that undergo apoptosis break apart and are recycled by a type of white blood cell called a macrophage (illustration). Apoptosis protects the body by removing genetically damaged cells that could lead to cancer, and it plays an important role in the development of the embryo and the maintenance of adult tissues.

Cancer results from a disruption of the normal regulation of the cell cycle. When the cycle proceeds without control, cells can divide without order and accumulate genetic defects that can lead to a cancerous tumor (illustration).

## How do geneticists indicate the location of a gene?

Geneticists use maps to describe the location of a particular gene on a chromosome. One type of map uses the cytogenetic location to describe a gene's position. The cytogenetic location is based on a distinctive pattern of bands created when chromosomes are stained with certain chemicals. Another type of map uses the molecular location, a precise description of a gene's position on a chromosome. The molecular location is based on the sequence of DNA building blocks (base pairs) that make up the chromosome.

## Cytogenetic Location

Geneticists use a standardised way of describing a gene's cytogenetic location. In most cases, the location describes the position of a particular band on a stained chromosome:

17q12

It can also be written as a range of bands, if less is known about the exact location:

17q12-q21

The combination of numbers and letters provide a gene's "address" on a chromosome. This address is made up of several parts:

- The chromosome on which the gene can be found. The first number or letter used to describe a gene's location represents the chromosome. Chromosomes 1 through 22 (the autosomes) are designated by their chromosome number. The sex chromosomes are designated by X or Y.
- The arm of the chromosome. Each chromosome is divided into two sections (arms) based on the location of a narrowing (constriction) called the centromere. By convention, the shorter arm is called p, and the longer arm is called q. The chromosome arm is the second part of the gene's address. For example, 5q is the long arm of chromosome 5, and Xp is the short arm of the X chromosome.
- The position of the gene on the p or q arm. The position of a gene is based on a distinctive pattern of light and dark bands that appear when the chromosome is stained in a certain way. The position is usually designated by two digits (representing a region and a band), which are sometimes followed by a decimal point and one or more additional digits (representing sub-bands within a light or dark area). The number indicating the gene position increases with distance from the centromere. For example: 14q21 represents position 21 on the long arm of chromosome 14. 14q21 is closer to the centromere than 14q22.

Sometimes, the abbreviations "cen" or "ter" are also used to describe a gene's cytogenetic location. "Cen" indicates that the gene is very close to the centromere. For example, 16pcen refers to the short arm of chromosome 16 near the centromere. "Ter" stands for terminus, which indicates that the gene is very close to the end of the p or q arm. For example, 14qter refers to the tip of the long arm of chromosome 14. ("Tel" is also sometimes used to describe a gene's location. "Tel" stands for telomeres, which are at the ends of each chromosome. The abbreviations "tel" and "ter" refer to the same location.)

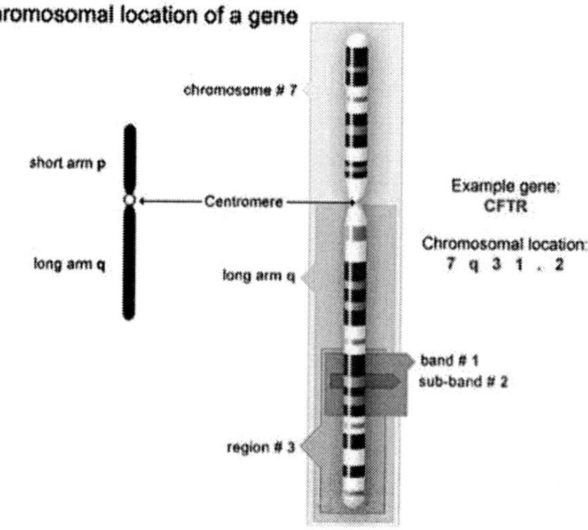


**Fig. 10.4:** The CFTR gene is located on the long arm of chromosome 7 at position 7q31.2

## Molecular Location

The Human Genome Project, an international research effort completed in 2003, determined the sequence of base pairs for each human chromosome. This sequence information allows researchers to provide a more specific address than the cytogenetic location for many genes. A gene's molecular address pinpoints the location of that gene in terms of base pairs. For example, the molecular location of the APOE gene on chromosome 19 begins with base pair 50,100,901 and ends with base pair 50,104,488. This range describes the gene's precise position on chromosome 19 and indicates the size of the gene (3,588 base pairs). Knowing a gene's molecular location also allows researchers to determine exactly how far the gene is from other genes on the same chromosome.

Different groups of researchers often present slightly different values for a gene's molecular location. Researchers interpret the sequence of the human genome using a variety of methods, which can result in small differences in a gene's molecular address. For example, the National Center for Biotechnology Information (NCBI) identifies the molecular location of the APOE gene as base pair 50,100,901 to base pair 50,104,488 on chromosome 19. The Ensembl database identifies the location of this gene as base pair 50,100,879 to base pair 50,104,489 on chromosome 19. Neither of these addresses is incorrect; they represent different interpretations of the same data. For consistency, Genetics Home Reference presents data from NCBI for the molecular location of genes.

### What are gene families?

A gene family is a group of genes that share important characteristics. In many cases, genes in a family share a similar sequence of DNA building blocks (nucleotides). These genes provide instructions for making products (such as proteins) that have a similar structure or function. In other cases, dissimilar genes are grouped together in a family because proteins produced from these genes work together as a unit or participate in the same process.

Classifying individual genes into families helps researchers describe how genes are related to each other. Researchers can use gene families to predict the function of newly identified genes based on their similarity to known genes. Similarities among genes in a family can also be used to predict where and when a specific gene is active (expressed). Additionally, gene families may provide clues for identifying genes that are involved in particular diseases.

Sometimes not enough is known about a gene to assign it to an established family. In other cases, genes may fit into more than one family. No formal guidelines define the criteria for grouping genes together. Classification systems for genes continue to evolve as scientists learn more about the structure and function of genes and the relationships between them.

## Mutations and Health

### What is a gene mutation and how do mutations occur?

A gene mutation is a permanent change in the DNA sequence that makes up a gene. Mutations range in size from a single DNA building block (DNA base) to a large segment of a chromosome.

Gene mutations occur in two ways: they can be inherited from a parent or acquired during a person's lifetime. Mutations that are passed from parent to child are called hereditary mutations or germline mutations (because they are present in the egg and sperm cells, which are also called germ cells). This type of mutation is present throughout a person's life in virtually every cell in the body.

Mutations that occur only in an egg or sperm cell, or those that occur just after fertilization, are called new (de novo) mutations. De novo mutations may explain genetic disorders in which an affected child has a mutation in every cell, but has no family history of the disorder.

Acquired (or somatic) mutations occur in the DNA of individual cells at some time during a person's life. These changes can be caused by environmental factors such as ultraviolet radiation from the sun, or can occur if a mistake is made as DNA copies itself during cell division. Acquired mutations in somatic cells (cells other than sperm and egg cells) cannot be passed on to the next generation.

Mutations may also occur in a single cell within an early embryo. As all the cells divide during growth and development, the individual will have some cells with the mutation and some cells without the genetic change. This situation is called mosaicism.

Some genetic changes are very rare; others are common in the population. Genetic changes that occur in more than 1 per cent of the population are called polymorphisms. They are common enough to be considered a normal variation in the DNA.

Polymorphisms are responsible for many of the normal differences between people such as eye colour, hair colour, and blood type. Although many polymorphisms have no negative effects on a person's health, some of these variations may influence the risk of developing certain disorders.

## How can gene mutations affect health and development?

To function correctly, each cell depends on thousands of proteins to do their jobs in the right places at the right times. Sometimes, gene mutations prevent one or more of these proteins from working properly. By changing a gene's instructions for making a protein, a mutation can cause the protein to malfunction or to be missing entirely. When a mutation alters a protein that plays a critical role in the body, it can disrupt normal development or cause a medical condition. A condition caused by mutations in one or more genes is called a genetic disorder.

In some cases, gene mutations are so severe that they prevent an embryo from surviving until birth. These changes occur in genes that are essential for development, and often disrupt the development of an embryo in its earliest stages. Because these mutations have very serious effects, they are incompatible with life.

It is important to note that genes themselves do not cause disease—genetic disorders are caused by mutations that make a gene function improperly. For example, when people say that someone has "the cystic fibrosis gene," they are usually referring to a mutated version of the CFTR gene, which causes the disease. All people, including those without cystic fibrosis, have a version of the CFTR gene.

## Do all gene mutations affect health and development?

No—only a small percentage of mutations cause genetic disorders—most have no impact on health or development. For example, some mutations alter a gene's DNA base sequence but do not change the function of the protein made by the gene.

Often, gene mutations that could cause a genetic disorder are repaired by certain enzymes before the gene is expressed (makes a protein). Each cell has a number of pathways through which enzymes recognise and repair mistakes in DNA. Because DNA can be damaged or mutated in many ways, DNA repair is an important process by which the body protects itself from disease.

A very small percentage of all mutations actually have a positive effect. These mutations lead to new versions of proteins that help an organism and its future generations better adapt to changes in their environment. For example, a beneficial mutation could result in a protein that protects the organism from a new strain of bacteria.

## What kinds of gene mutations are possible?

The DNA sequence of a gene can be altered in a number of ways. Gene mutations have varying effects on health, depending on where they occur and whether they alter the function of essential proteins. The types of mutations include:

***Missense Mutation:*** This type of mutation is a change in one DNA base pair that results in the substitution of one amino acid for another in the protein made by a gene.

***Nonsense Mutation:*** A nonsense mutation is also a change in one DNA base pair. Instead of substituting one amino acid for another, however, the altered DNA sequence prematurely signals the cell to stop building a protein. This type of mutation results in a shortened protein that may function improperly or not at all.

***Insertion:*** An insertion changes the number of DNA bases in a gene by adding a piece of DNA. As a result, the protein made by the gene may not function properly.

***Deletion***: A deletion changes the number of DNA bases by removing a piece of DNA. Small deletions may remove one or a few base pairs within a gene, while larger deletions can remove an entire gene or several neighbouring genes. The deleted DNA may alter the function of the resulting protein(s).

***Duplication:*** A duplication consists of a piece of DNA that is abnormally copied one or more times. This type of mutation may alter the function of the resulting protein.

***Frameshift Mutation***: This type of mutation occurs when the addition or loss of DNA bases changes a gene's reading frame. A reading frame consists of groups of 3 bases that each code for one amino acid. A frameshift mutation shifts the grouping of these bases and changes the code for amino acids. The resulting protein is usually nonfunctional. Insertions, deletions, and duplications can all be frameshift mutations.

***Repeat Expansion***: Nucleotide repeats are short DNA sequences that are repeated a number of times in a row. For example, a trinucleotide repeat is made up of 3-base-pair sequences, and a tetranucleotide repeat is made up of 4-base-pair sequences. A repeat expansion is a mutation that increases the number of times that the short DNA sequence is repeated. This type of mutation can cause the resulting protein to function improperly.

## Can a change in the number of genes affect health and development?

People have two copies of most genes, one copy inherited from each parent. In some cases, however, the number of copies varies—meaning that a person can be born with one, three, or more copies of particular genes. Less commonly, one or more genes may be entirely missing. This type of genetic difference is known as copy number variation (CNV).

Copy number variation results from insertions, deletions, and duplications of large segments of DNA. These segments are big enough to include whole genes. Variation in gene copy number can influence the activity of genes and ultimately affect many body functions.

Researchers were surprised to learn that copy number variation accounts for a significant amount of genetic difference between people. More than 10 per cent of human DNA appears to contain these differences in gene copy number. While much of this variation does not affect health or development, some differences likely influence a person's risk of disease and response to certain drugs. Future research will focus on the consequences of copy number variation in different parts of the genome and study the contribution of these variations to many types of disease.

## Can changes in the number of chromosomes affect health and development?

Human cells normally contain 23 pairs of chromosomes, for a total of 46 chromosomes in each cell. A change in the number of chromosomes can cause problems with growth, development, and function of the body's systems. These changes can occur during the formation of reproductive cells (eggs and sperm), in early fetal development, or in any cell after birth. A gain or loss of chromosomes from the normal 46 is called aneuploidy.

A common form of aneuploidy is trisomy, or the presence of an extra chromosome in cells. "Tri-" is Greek for "three"; people with trisomy have three copies of a particular chromosome in cells instead of the normal two copies. Down syndrome is an example of a condition caused by trisomy. People with Down syndrome typically have three copies of chromosome 21 in each cell, for a total of 47 chromosomes per cell.

Monosomy, or the loss of one chromosome in cells, is another kind of aneuploidy. "Mono-" is Greek for "one"; people with monosomy have one copy of a particular chromosome in cells instead of the normal two copies. Turner syndrome is a condition caused by monosomy. Women with Turner syndrome usually have only one copy of the X chromosome in every cell, for a total of 45 chromosomes per cell.

Rarely, some cells end up with complete extra sets of chromosomes. Cells with one additional set of chromosomes, for a total of 69 chromosomes, are called triploid. Cells with two additional sets of chromosomes, for a total of 92 chromosomes, are called tetraploid. A condition in which every cell in the body has an extra set of chromosomes is not compatible with life.

In some cases, a change in the number of chromosomes occurs only in certain cells. When an individual has two or more cell populations with a different chromosomal make up, this situation is called chromosomal mosaicism. Chromosomal mosaicism occurs from an error in cell division in cells other than eggs and sperm. Most commonly, some cells end up with one extra or missing chromosome (for a total of 45 or 47 chromosomes per cell), while other cells have the usual 46 chromosomes. Mosaic

Turner syndrome is one example of chromosomal mosaicism. In females with this condition, some cells have 45 chromosomes because they are missing one copy of the X chromosome, while other cells have the usual number of chromosomes.

Many cancer cells also have changes in their number of chromosomes. These changes are not inherited; they occur in somatic cells (cells other than eggs or sperm) during the formation or progression of a cancerous tumor.

### Can changes in the structure of chromosomes affect health and development?

Changes that affect the structure of chromosomes can cause problems with growth, development, and function of the body's systems. These changes can affect many genes along the chromosome and disrupt the proteins made from those genes.

Structural changes can occur during the formation of egg or sperm cells, in early fetal development, or in any cell after birth. Pieces of DNA can be rearranged within one chromosome or transferred between two or more chromosomes. The effects of structural changes depend on their size and location, and whether any genetic material is gained or lost. Some changes cause medical problems, while others may have no effect on a person's health.

### Changes in chromosome structure include:

***Translocations:*** A translocation occurs when a piece of one chromosome breaks off and attaches to another chromosome. This type of rearrangement is described as balanced if no genetic material is gained or lost in the cell. If there is a gain or loss of genetic material, the translocation is described as unbalanced.

***Deletions:*** Deletions occur when a chromosome breaks and some genetic material is lost. Deletions can be large or small, and can occur anywhere along a chromosome.

***Duplications***: Duplications occur when part of a chromosome is copied (duplicated) too many times. This type of chromosomal change results in extra copies of genetic material from the duplicated segment.

***Inversions:*** An inversion involves the breakage of a chromosome in two places: the resulting piece of DNA is reversed and re-inserted into the chromosome. Genetic material may or may not be lost as a result of the chromosome breaks. An inversion that involves the chromosome's constriction point (centromere) is called a pericentric inversion. An inversion that occurs in the long (q) arm or short (p) arm and does not involve the centromere is called a paracentric inversion.

***Isochromosomes:*** An isochromosome is a chromosome with two identical arms. Instead of one long (q) arm and one short (p) arm, an isochromosome has two long arms or two short arms. As a result, these abnormal chromosomes have an extra copy of some genes and are missing copies of other genes.

***Dicentric Chromosomes***: Unlike normal chromosomes, which have a single constriction point (centromere), a dicentric chromosome contains two centromeres. Dicentric chromosomes result from the abnormal fusion of two chromosome pieces, each of which includes a centromere. These structures are unstable and often involve a loss of some genetic material.

***Ring Chromosomes***: Ring chromosomes usually occur when a chromosome breaks in two places and the ends of the chromosome arms fuse together to form a circular structure. The ring may or may not include the chromosome's constriction point (centromere). In many cases, genetic material near the ends of the chromosome is lost.

Many cancer cells also have changes in their chromosome structure. These changes are not inherited; they occur in somatic cells (cells other than eggs or sperm) during the formation or progression of a cancerous tumour.

## Can changes in mitochondrial DNA affect health and development?

Mitochondria are structures within cells that convert the energy from food into a form that cells can use. Although most DNA is packaged in chromosomes within the nucleus, mitochondria also have a small amount of their own DNA (known as mitochondrial DNA or mtDNA). In some cases, inherited changes in mitochondrial DNA can cause problems with growth, development, and function of the body's systems. These mutations disrupt the mitochondria's ability to generate energy efficiently for the cell.

Conditions caused by mutations in mitochondrial DNA often involve multiple organ systems. The effects of these conditions are most pronounced in organs and tissues that require a lot of energy (such as the heart, brain, and muscles). Although the health consequences of inherited mitochondrial DNA mutations vary widely, frequently observed features include muscle weakness and wasting, problems with movement, diabetes, kidney failure, heart disease, loss of intellectual functions (dementia), hearing loss, and abnormalities involving the eyes and vision.

Mitochondrial DNA is also prone to noninherited (somatic) mutations. Somatic mutations occur in the DNA of certain cells during a person's lifetime, and typically are not passed to future generations. Because mitochondrial DNA has a limited ability to repair itself when it is damaged, these mutations tend to build up over time. A buildup of somatic mutations in mitochondrial DNA has been associated with some forms of cancer and an increased risk of certain age-related disorders such as heart disease, Alzheimer disease, and Parkinson disease. Additionally, research suggests that the progressive accumulation of these mutations over a person's lifetime may play a role in the normal process of aging.

## What are complex or multifactorial disorders?

Researchers are learning that nearly all conditions and diseases have a genetic component. Some disorders, such as sickle cell anaemia and cystic fibrosis, are caused by mutations in a single gene. The causes of many other disorders, however, are much more complex. Common medical problems such as heart disease, diabetes, and obesity do not have a single genetic cause—they are likely associated with the effects of multiple genes in combination with lifestyle and environmental factors. Conditions caused by many contributing factors are called complex or multifactorial disorders.

Although complex disorders often cluster in families, they do not have a clear-cut pattern of inheritance. This makes it difficult to determine a person's risk of inheriting or passing on these disorders. Complex disorders are also difficult to study and treat because the specific factors that cause most of these disorders have not yet been identified. By 2010, however, researchers predict they will have found the major contributing genes for many common complex disorders.

## How are genetic conditions and genes named?

*Naming Genetic Conditions:* Genetic conditions are not named in one standard way (unlike genes, which are given an official name and symbol by a formal committee). Doctors who treat families with a particular disorder are often the first to propose a name for the condition. Expert working groups may later revise the name to improve its usefulness. Naming is important because it allows accurate and effective communication about particular conditions, which will ultimately help researchers find new approaches to treatment.

Disorder names are often derived from one or a combination of sources:

- The basic genetic or biochemical defect that causes the condition (for example, alpha-1 antitrypsin deficiency);
- One or more major signs or symptoms of the disorder (for example, sickle cell anaemia);
- The parts of the body affected by the condition (for example, retinoblastoma);
- The name of a physician or researcher, often the first person to describe the disorder (for example, Marfan syndrome, which was named after Dr. Antoine Bernard-Jean Marfan);
- A geographic area (for example, familial Mediterranean fever, which occurs mainly in populations bordering the Mediterranean Sea); or
- The name of a patient or family with the condition (for example, amyotrophic lateral sclerosis, which is also called Lou Gehrig disease after a famous baseball player who had the condition).

Disorders named after a specific person or place are called eponyms. There is debate as to whether the possessive form (e.g., Alzheimer's disease) or the nonpossessive form (Alzheimer disease) of eponyms is preferred. As a rule, medical geneticists use the nonpossessive form, and this form may become the standard for doctors in all fields of medicine.

## Naming Genes

The HUGO Gene Nomenclature Committee. This link leads to a site outside Genetics Home Reference. (HGNC) designates an official name and symbol (an abbreviation of the name) for each known human gene. Some official gene names include additional information in parentheses, such as related genetic conditions, subtypes of a condition, or inheritance pattern. The HGNC is a non-profit organisation funded by the U.K. Medical Research Council and the U.S. National Institutes of Health. The Committee has named more than 13,000 of the estimated 20,000 to 25,000 genes in the human genome.

During the research process, genes often acquire several alternate names and symbols. Different researchers investigating the same gene may each give the gene a different name, which can cause confusion. The HGNC assigns a unique name and symbol to each human gene, which allows effective organisation of genes in large databanks, aiding the advancement of research. For specific information about how genes are named, refer to the HGNC's Guidelines for Human Gene Nomenclature. This link leads to a site outside Genetics Home Reference.

# 11

# Inheritance of Genetic Conditions

## Introduction

A genetic condition can be autosomal or X-linked, depending on the location of the gene. It can also be dominant or recessive, depending on the function of the gene.

The following explains the 3 kinds of inherited genetic conditions:

- Dominant inheritance
- Recessive inheritance
- X-linked inheritance

## Dominant Inheritance

In dominant inheritance, a disease is caused if one gene of a pair does not function well or at all. In the human body, one poorly or non-functioning gene will cause damage, which will result in a genetic condition.

Examples for genetic conditions inherited in this manner are Huntington disease and many types of dwarfism.

Since you pass on to your child one or the other gene of a pair the baby has 1 in 2 chance for inheriting the affected or normal gene. Thus the risk for a parent with an autosomal dominant condition to pass the disease to his/her child is 50 per cent in each pregnancy. Autosomal dominant condition (the problem gene resides on one of the 22 chromosome pairs which both boys and girls have) can affected both sexes.

It does not matter how many brothers or sisters have the disease.

Occasionally a new dominant trait emerges in a family: a child is born with certain condition, known to have a dominant inheritance, which neither parents has. This is called a new dominant mutation, which can occur, in the specific egg or sperm, which conceived the child. The recurrence risk for the parents will be low but for the affected child, in her/his offspring, it will be 50 per cent. The reasons for such a new mutation to emerge is not known. It is not caused by anything that the parents did or did not do. In some cases, association with older father's age was noted.

Why is recurrence risk for parents of such an affected child not the same as the risk for the general population? In some cases, more than one affected child was born to such a couple due to a condition called gonadal mosaicism.

Gonadal mosaicism means that more than one egg or sperm carrying this mutation exist in the testes or ovaries of a parent. The exact recurrence risk in such cases cannot be calculated. It has to be obtained through a population study. This means that we have to contact all couples who had a child with this condition but the parents themselves were unaffected. From this we can find out how many couples had more than one affected child. A risk calculated in this manner an empiric risk.

## Recessive Inheritance

Recessive genes, unlike dominant genes, only show up in a person when both genes of a pair are changed. If a person has a change in one gene of a pair he/she is called a carrier of a recessive gene and will not have any clinical manifestations. Recessive genes cause many diseases.

The most common diseases are: Cystic fibrosis in the Caucasian population; Sickle cell anaemia in the black population; Tay-Sachs disease in the Ashkenazi-Jewish population; Alpha thalassaemia in the oriental population.

Everybody carries 6-8 recessive genes, which do not function well. So, we are all carriers of some recessive conditions without knowing it. Fortunately, only one gene of a pair is mutated while the other is normal and dominates the problem geneand therefore doesn't affect the individual.

It is only when two parents who carry the same recessive problem gene have children that the child can inherit the recessive problem gene from each of the carrier parents. As a result, the child has a pair of recessive problem genes and will thus have the condition.

This can happen by chance or in consanguineous marriage (first cousins for example) or when both parents belong to the same ethnic background.

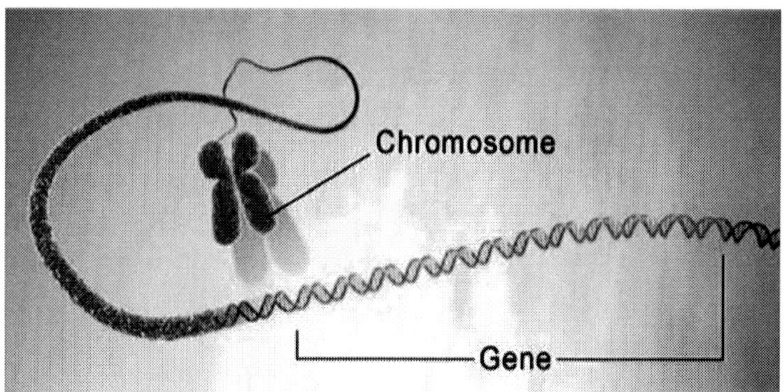


**Fig. 11.1:** Genes are made up of DNA. Each chromosome contains many genes

As shown in the diagram above, each parent gives a baby one gene from a pair. So there are four possible combinations of the parents' genes and the baby will inherit one of these combinations:

## X-linked Inheritance

The mother has two X-chromosomes and thus always gives X-chromosome to her offspring. The father has one X- and one Y-chromosome and can give either chromosome to his offspring.

If the Y-chromosome is inherited from the father, then the baby is a boy. If the X-chromosome is inherited from the father, then the baby is a girl. Thus, the sperm of the father determines the sex of the baby and there is 50-50 chance as to whether a X- or Y-chromosome will be passed along.

The X-chromosome has many genes, many which are involved in the function of many body organs.

The Y-chromosome is small and contains a small number of genes, most of which are involved in determining the male sex.

Since a girl has two X-chromosome the genes on them all have pairs, or partners. If there is a problem gene on the X-chromosome, it is usually paired with a normal gene which covers up the problem.

However, a boy has only one X-chromosome, so the genes on it don't have partners. Thus, problem genes on the X-chromosomes, don't have partners to cover up the problem and the condition manifests.

In most cases, girls and women with a problem gene on an X-chromosome are carriers of X-linked conditions, unaffected (and usually unaware) by the genetic condition.

Boys and men, with the X-linked problem gene will have the trait.

Therefore, most X-linked conditions are X-linked recessive. That is, they do not show any problems in girls and women, only in boys and men.

Examples of X-linked genetic traits are red-green color blindness, Fragile X syndrome, hemophilia and Duchenne muscular dystrophy. A boy with these conditions inherits them from his mother's X-chromosome.

The *Fig. 11.1* shows the four possible gene combinations for a child of a mother who is a carrier for a serious X-linked disease and a father whose X-chromosome contains the normal gene (and is thus unaffected).

Thus, when an X-linked condition is carried by the mother, each baby girl has 50 per cent chance for being a carrier. Each baby boy has 50 per cent chance of having the condition itself.

As in autosomal dominant conditions, the change in an X-linked gene which turns it into a problem gene can happen in the specific egg that conceived the boy or the specific egg or sperm that conceived the girl. In fact, in many cases of hemophilia, the change in the X-linked gene started in the sperm that conceived the mother (in the maternal grandfather) of the affected boy.

There are also some X-linked dominant conditions. That means a problem gene on the X-chromosome that causes problems in a girl who carries the problem gene on one of her X-chromosomes. These rare, severe, many times fatal conditions in boys because they only have one X-chromosome.

## What does it mean if a disorder seems to run in family?

A particular disorder might be described as "running in a family" if more than one person in the family has the condition. Some disorders that affect multiple family members are caused by gene mutations, which can be inherited (passed down from parent to child). Other conditions that appear to run in families are not caused by mutations in single genes. Instead, environmental factors such as dietary habits or a combination of genetic and environmental factors are responsible for these disorders.

It is not always easy to determine whether a condition in a family is inherited. A genetics professional can use a person's family history (a record of health information about a person's immediate and extended family) to help determine whether a disorder has a genetic component. He or she will ask about the health of people from several generations of the family, usually first-, second-, and third-degree relatives.

## Why is it important to know family medical history?

A family medical history is a record of health information about a person and his or her close relatives. A complete record includes information from three generations of relatives, including children, brothers and sisters, parents, aunts and uncles, nieces and nephews, grandparents, and cousins.

Families have many factors in common, including their genes, environment, and lifestyle. Together, these factors can give clues to medical conditions that may run in a family. By noticing patterns of disorders among relatives, healthcare professionals can determine whether an individual, other family members, or future generations may be at an increased risk of developing a particular condition.

A family medical history can identify people with a higher-than-usual chance of having common disorders, such as heart disease, high blood pressure, stroke, certain cancers, and diabetes. These complex disorders are influenced by a combination of genetic factors, environmental conditions, and lifestyle choices. A family history also can provide information about the risk of rarer conditions caused by mutations in a single gene, such as cystic fibrosis and sickle cell anaemia.

While a family medical history provides information about the risk of specific health concerns, having relatives with a medical condition does not mean that an individual will definitely develop that condition. On the other hand, a person with no family history of a disorder may still be at risk of developing that disorder.

Knowing one's family medical history allows a person to take steps to reduce his or her risk. For people at an increased risk of certain cancers, healthcare professionals may recommend more frequent screening (such as mammography or colonoscopy) starting at an earlier age. Healthcare providers may also encourage regular checkups or testing for people with a medical condition that runs in their family. Additionally, lifestyle changes such as adopting a healthier diet, getting regular exercise, and quitting smoking help many people lower their chances of developing heart disease and other common illnesses.

The easiest way to get information about family medical history is to talk to relatives about their health. Have they had any medical problems, and when did they occur? A family gathering could be a good time to discuss these issues. Additionally, obtaining medical records and other documents (such as obituaries and death certificates) can help complete a family medical history. It is important to keep this information up-to-date and to share it with a healthcare professional regularly.

## If a genetic disorder runs in my family

### what are the chances that my children will have the condition?

When a genetic disorder is diagnosed in a family, family members often want to know the likelihood that they or their children will develop the condition. This can be difficult to predict in some cases because many factors influence a person's chances of developing a genetic condition. One important factor is how the condition is inherited. For example:

Autosomal dominant inheritance: A person affected by an autosomal dominant disorder has a 50 per cent chance of passing the mutated gene to each child. The chance that a child will not inherit the mutated gene is also 50 per cent.

Autosomal recessive inheritance: Two unaffected people who each carry one copy of the mutated gene for an autosomal recessive disorder (carriers) have a 25 per cent chance with each pregnancy of having a child affected by the disorder. The chance with each pregnancy of having an unaffected child who is a carrier of the disorder is 50 per cent, and the chance that a child will not have the disorder and will not be a carrier is 25 per cent.

X-linked dominant inheritance: The chance of passing on an X-linked dominant condition differs between men and women because men have one X chromosome and one Y chromosome, while women have two X chromosomes. A man passes on his Y chromosome to all of his sons and his X chromosome to all of his daughters. Therefore, the sons of a man with an X-linked dominant disorder will not be affected, but all of his daughters will inherit the condition (illustration). A woman passes on one or the other of her X chromosomes to each child. Therefore, a woman with an X-linked dominant disorder has a 50 per cent chance of having an affected daughter or son with each pregnancy.

X-linked recessive inheritance: Because of the difference in sex chromosomes, the probability of passing on an X-linked recessive disorder also differs between men and women. The sons of a man with an X-linked recessive disorder will not be affected, and his daughters will carry one copy of the mutated gene. With each pregnancy, a woman who carries an X-linked recessive disorder has a 50 percent chance of having sons who are affected and a 50 percent chance of having daughters who carry one copy of the mutated gene.

Codominant inheritance: In codominant inheritance, each parent contributes a different version of a particular gene, and both versions influence the resulting genetic trait. The chance of developing a genetic condition with codominant inheritance, and the characteristic features of that condition, depend on which versions of the gene are passed from parents to their child.

Mitochondrial inheritance: Mitochondria, which are the energy-producing centers inside cells, each contain a small amount of DNA. Disorders with mitochondrial inheritance result from mutations in mitochondrial DNA. Although mitochondrial disorders can affect both males and females, only females can pass mutations in mitochondrial DNA to their children. A woman with a disorder caused by changes in mitochondrial DNA will pass the mutation to all of her daughters and sons, but the children of a man with such a disorder will not inherit the mutation.

It is important to note that the chance of passing on a genetic condition applies equally to each pregnancy. For example, if a couple has a child with an autosomal recessive disorder, the chance of having another child with the disorder is still 25 per cent (or 1 in 4). Having one child with a disorder does not "protect" future children from inheriting the condition. Conversely, having a child without the condition does not mean that future children will definitely be affected.

Although the chances of inheriting a genetic condition appear straight forward, factors such as a person's family history and the results of genetic testing can sometimes modify those chances. In addition, some people with a disease-causing mutation never develop any health problems or may experience only mild symptoms of the disorder. If a disease that runs in a family does not have a clear-cut inheritance pattern, predicting the likelihood that a person will develop the condition can be particularly difficult.

Estimating the chance of developing or passing on a genetic disorder can be complex. Genetics professionals can help people understand these chances and help them make informed decisions about their health.

## What are reduced penetrance and variable expressivity?

Reduced penetrance and variable expressivity are factors that influence the effects of particular genetic changes. These factors usually affect disorders that have an autosomal dominant pattern of inheritance, although they are occasionally seen in disorders with an autosomal recessive inheritance pattern.

***Reduced Penetrance:*** Penetrance refers to the proportion of people with a particular genetic change (such as a mutation in a specific gene) who exhibit signs and symptoms of a genetic disorder. If some people with the mutation do not develop features of the disorder, the condition is said to have reduced (or incomplete) penetrance. Reduced penetrance often occurs with familial cancer syndromes. For example, many people with a mutation in the BRCA1 or BRCA2 gene will develop cancer during their lifetime, but some people will not. Doctors cannot predict which people with these mutations will develop cancer or when the tumors will develop.

Reduced penetrance probably results from a combination of genetic, environmental, and lifestyle factors, many of which are unknown. This phenomenon can make it challenging for genetics professionals to interpret a person's family medical history and predict the risk of passing a genetic condition to future generations.

***Variable Expressivity:*** Although some genetic disorders exhibit little variation, most have signs and symptoms that differ among affected individuals. Variable expressivity refers to the range of signs and symptoms that can occur in different people with the same genetic condition. For example, the features of Marfan syndrome vary widely— some people have only mild symptoms (such as being tall and thin with long, slender fingers), while others also experience life-threatening complications involving the heart and blood vessels. Although the features are highly variable, most people with this disorder have a mutation in the same gene (FBN1).

As with reduced penetrance, variable expressivity is probably caused by a combination of genetic, environmental, and lifestyle factors, most of which have not been identified. If a genetic condition has highly variable signs and symptoms, it may be challenging to diagnose.

## What do Geneticists mean by Anticipation?

The signs and symptoms of some genetic conditions tend to become more severe and appear at an earlier age as the disorder is passed from one generation to the next. This phenomenon is called anticipation. Anticipation is most often seen with certain genetic disorders of the nervous system, such as Huntington disease, myotonic dystrophy, and fragile X syndrome.

Anticipation typically occurs with disorders that are caused by an unusual type of mutation called a trinucleotide repeat expansion. A trinucleotide repeat is a sequence of three DNA building blocks (nucleotides) that is repeated a number of times in a row. DNA segments with an abnormal number of these repeats are unstable and prone to errors during cell division. The number of repeats can change as the gene is passed from parent to child. If the number of repeats increases, it is known as a trinucleotide repeat expansion. In some cases, the trinucleotide repeat may expand until the gene stops functioning normally. This expansion causes the features of some disorders to become more severe with each successive generation.

Most genetic disorders have signs and symptoms that differ among affected individuals, including affected people in the same family. Not all of these differences can be explained by anticipation. A combination of genetic, environmental, and lifestyle factors is probably responsible for the variability, although many of these factors have not been identified. Researchers study multiple generations of affected family members and consider the genetic cause of a disorder before determining that it shows anticipation.

## What are Genomic Imprinting and Uniparental Disomy?

Genomic imprinting and uniparental disomy are factors that influence how some genetic conditions are inherited.

*Genomic Imprinting:* People inherit two copies of their genes—one from their mother and one from their father. Usually both copies of each gene are active, or "turned on," in cells. In some cases, however, only one of the two copies is normally turned on. Which copy is active depends on the parent of origin: some genes are normally active only when they are inherited from a person's father; others are active only when inherited from a person's mother. This phenomenon is known as genomic imprinting.

In genes that undergo genomic imprinting, the parent of origin is often marked, or "stamped," on the gene during the formation of egg and sperm cells. This stamping process, called methylation, is a chemical reaction that attaches small molecules called methyl groups to certain segments of DNA. These molecules identify which copy of a gene was inherited from the mother and which was inherited from the father. The addition and removal of methyl groups can be used to control the activity of genes.

Only a small percentage of all human genes undergo genomic imprinting. Researchers are not yet certain why some genes are imprinted and others are not. They do know that imprinted genes tend to cluster together in the same regions of chromosomes. Two major clusters of imprinted genes have been identified in humans, one on the short (p) arm of chromosome 11 (at position 11p15) and another on the long (q) arm of chromosome 15 (in the region 15q11 to 15q13).

*Uniparental Disomy*: Uniparental disomy (UPD) occurs when a person receives two copies of a chromosome, or part of a chromosome, from one parent and no copies from the other parent. UPD can occur as a random event during the formation of egg or sperm cells or may happen in early fetal development.

In many cases, UPD likely has no effect on health or development. Because most genes are not imprinted, it doesn't matter if a person inherits both copies from one parent instead of one copy from each parent. In some cases, however, it does make a difference whether a gene is inherited from a person's mother or father. A person with UPD may lack any active copies of essential genes that undergo genomic imprinting. This loss of gene function can lead to delayed development, mental retardation, or other medical problems.

Several genetic disorders can result from UPD or a disruption of normal genomic imprinting. The most well-known conditions include Prader-Willi syndrome, which is characterised by uncontrolled eating and obesity, and Angelman syndrome, which causes mental retardation and impaired speech. Both of these disorders can be caused by UPD or other errors in imprinting involving genes on the long arm of chromosome 15. Other conditions, such as Beckwith-Wiedemann syndrome (a disorder characterised by accelerated growth and an increased risk of cancerous tumors), are associated with abnormalities of imprinted genes on the short arm of chromosome 11.

## Are Chromosomal Disorders Inherited?

Although it is possible to inherit some types of chromosomal abnormalities, most chromosomal disorders (such as Down syndrome and Turner syndrome) are not passed from one generation to the next.

Some chromosomal conditions are caused by changes in the number of chromosomes. These changes are not inherited, but occur as random events during the formation of reproductive cells (eggs and sperm). An error in cell division called non-disjunction results in reproductive cells with an abnormal number of chromosomes. For example, a reproductive cell may accidentally gain or lose one copy of a chromosome. If one of these atypical reproductive cells contributes to the genetic make up of a child, the child will have an extra or missing chromosome in each of the body's cells.

Changes in chromosome structure can also cause chromosomal disorders. Some changes in chromosome structure can be inherited, while others occur as random accidents during the formation of reproductive cells or in early fetal development.

Because the inheritance of these changes can be complex, people concerned about this type of chromosomal abnormality may want to talk with a genetics professional.

Some cancer cells also have changes in the number or structure of their chromosomes. Because these changes occur in somatic cells (cells other than eggs and sperm), they cannot be passed from one generation to the next.

Why are some genetic conditions more common in particular ethnic groups?

Some genetic disorders are more likely to occur among people who trace their ancestry to a particular geographic area. People in an ethnic group often share certain versions of their genes, which have been passed down from common ancestors. If one of these shared genes contains a disease-causing mutation, a particular genetic disorder may be more frequently seen in the group.

Examples of genetic conditions that are more common in particular ethnic groups are sickle cell anaemia, which is more common in people of African, African-American, or Mediterranean heritage; and Tay-Sachs disease, which is more likely to occur among people of Ashkenazi (eastern and central European) Jewish or French Canadian ancestry. It is important to note, however, that these disorders can occur in any ethnic group.

## Genetic Consultation

### What is a Genetic Consultation?

A genetic consultation is a health service that provides information and support to people who have, or may be at risk for, genetic disorders. During a consultation, a genetics professional meets with an individual or family to discuss genetic risks or to diagnose, confirm, or rule out a genetic condition.

Genetics professionals include medical geneticists (doctors who specialise in genetics) and genetic counselors (certified healthcare workers with experience in medical genetics and counseling). Other healthcare professionals such as nurses, psychologists, and social workers trained in genetics can also provide genetic consultations.

Consultations usually take place in a doctor's office, hospital, genetics center, or other type of medical center. These meetings are most often in-person visits with individuals or families, but they are occasionally conducted in a group or over the telephone.

## Why might someone have a genetic consultation?

Individuals or families who are concerned about an inherited condition may benefit from a genetic consultation. The reasons that a person might be referred to a genetic counselor, medical geneticist, or other genetics professional include:

A personal or family history of a genetic condition, birth defect, chromosomal disorder, or hereditary cancer.

Two or more pregnancy losses (miscarriages), a stillbirth, or a baby who died.

A child with a known inherited disorder, a birth defect, mental retardation, or developmental delay.

A woman who is pregnant or plans to become pregnant at or after age 35. (Some chromosomal disorders occur more frequently in children born to older women.)

Abnormal test results that suggest a genetic or chromosomal condition.

An increased risk of developing or passing on a particular genetic disorder on the basis of a person's ethnic background.

People related by blood (for example, cousins) who plan to have children together. (A child whose parents are related may be at an increased risk of inheriting certain genetic disorders.)

A genetic consultation is also an important part of the decision-making process for genetic testing. A visit with a genetics professional may be helpful even if testing is not available for a specific condition, however.

## What happens during a genetic consultation?

A genetic consultation provides information, offers support, and addresses a patient's specific questions and concerns. To help determine whether a condition has a genetic component, a genetics professional asks about a person's medical history and takes a detailed family history (a record of health information about a person's immediate and extended family). The genetics professional may also perform a physical examination and recommend appropriate tests.

If a person is diagnosed with a genetic condition, the genetics professional provides information about the diagnosis, how the condition is inherited, the chance of passing the condition to future generations, and the options for testing and treatment.

During a consultation, a genetics professional will:

Interpret and communicate complex medical information.

Help each person make informed, independent decisions about their health care and reproductive options.

Respect each person's individual beliefs, traditions, and feelings.

A genetics professional will NOT:

Tell a person which decision to make.

Advise a couple not to have children.

Recommend that a woman continue or end a pregnancy.

Tell someone whether to undergo testing for a genetic disorder.

## How can I find a genetics professional in my area?

To find a genetics professional in your community, you may wish to ask your doctor for a referral. If you have health insurance, you can also contact your insurance company to find a medical geneticist or genetic counselor in your area who participates in your plan.

Several resources for locating a genetics professional in your community are available online:

GeneTests from the University of Washington provides a list of genetics clinics around the United StatesThis link leads to a site outside Genetics Home Reference. and international genetics clinics. This link leads to a site outside Genetics Home Reference.. You can also access the list by clicking on "Clinic Directory" at the top of the GeneTests home page. Clinics can be chosen by state or country, by service, and/or by specialty. State maps can help you locate a clinic in your area.

The National Society of Genetic Counselors offers a searchable directory of genetic counselors in the United StatesThis link leads to a site outside Genetics Home Reference. You can search by location, name, area of practice/specialisation, and/or ZIP Code.

The National Cancer Institute provides a Cancer Genetics Services Directory. This link leads to a site outside Genetics Home Reference., which lists professionals who provide services related to cancer genetics. You can search by type of cancer or syndrome, location, and/or provider name.

# 12

# Mitochondrial DNA

## Introduction

Mitochondria are structures within cells that convert the energy from food into a form that cells can use. Although most DNA is packaged in chromosomes within the nucleus, mitochondria also have a small amount of their own DNA. This genetic material is known as mitochondrial DNA or mtDNA. In humans, mitochondrial DNA spans about 16,500 DNA building blocks (base pairs), representing a fraction of the total DNA in cells.

Mitochondrial DNA contains 37 genes, all of which are essential for normal mitochondrial function. Thirteen of these genes provide instructions for making enzymes involved in oxidative phosphorylation. Oxidative phosphorylation is a process that uses oxygen and simple sugars to create adenosine triphosphate (ATP), the cell's main energy source. The remaining genes provide instructions for making molecules called transfer RNAs (tRNAs) and ribosomal RNAs (rRNAs), which are chemical cousins of DNA. These types of RNA help assemble protein building blocks (amino acids) into functioning proteins.

Mitochondrial genes are among the estimated 20,000 to 25,000 total genes in the human genome.

There are genetic conditions related to mitochondrial genes.

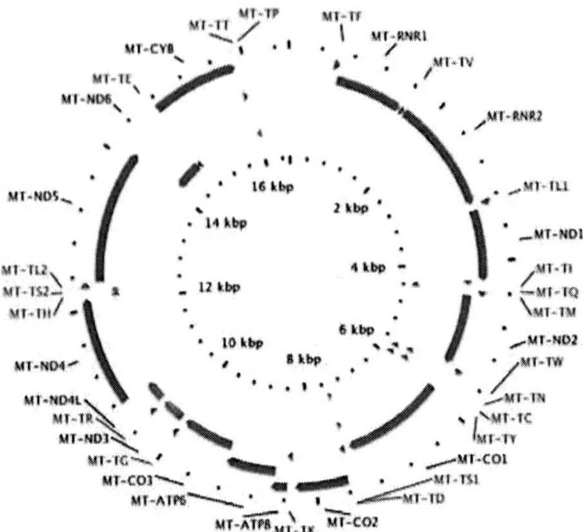


**Fig. 12.1:** Mitochondrial DNA is typically diagrammed as a circular structure with genes and regularly regions labelled.

## What conditions are related to mitochondrial DNA?

The following conditions are related to changes in mitochondrial DNA.

### Cancers

Mitochondrial DNA is prone to non-inherited (somatic) mutations. Somatic mutations occur in the DNA of certain cells during a person's lifetime and typically are not passed to future generations. Somatic mutations in mitochondrial DNA have been reported in some forms of cancer, including breast, colon, stomach, liver, and kidney tumors. These mutations also have been associated with cancer of blood-forming tissue (leukaemia) and cancer of immune system cells (lymphoma).

Somatic mutations in mitochondrial DNA may increase the production of potentially harmful molecules called reactive oxygen species. Mitochondrial DNA is particularly vulnerable to the effects of these molecules and has a limited ability to repair itself. As a result, reactive oxygen species easily damage mitochondrial DNA, causing a buildup of additional somatic mutations. Researchers continue to investigate how these mutations may lead to uncontrolled cell division and the growth of cancerous tumours.

## Leber Hereditary Optic Neuropathy

Mutations in four mitochondrial genes, MT-ND1, MT-ND4, MT-ND4L, and MT-ND6, have been identified in people with Leber hereditary optic neuropathy. These genes provide instructions for making proteins that are part of a large enzyme complex. This group of enzymes, known as Complex I, is necessary for oxidative phosphorylation. The mutations responsible for Leber hereditary optic neuropathy change single protein building blocks (amino acids) in these proteins, which may affect the generation of ATP within mitochondria. It remains unclear, however, why the effects of these mutations are often limited to the nerve that relays visual information from the eye to the brain (the optic nerve). Additional genetic and environmental factors probably contribute to the vision loss and other medical problems associated with Leber hereditary optic neuropathy.

## Mitochondrial Encephalomyopathy, Lactic Acidosis, and Stroke-like Episodes

Mutations in at least five mitochondrial genes, MT-ND1, MT-ND5, MT-TH, MT-TL1, and MT-TV, can cause the characteristic features of mitochondrial encephalopathy, lactic acidosis, and stroke-like episodes (MELAS). Some of these genes provide instructions for making proteins that are part of a large enzyme complex, called Complex I, that is necessary for oxidative phosphorylation. The other genes provide instructions for making transfer RNA molecules, which are essential for protein production within mitochondria.

Mutations in one transfer RNA gene, MT-TL1, cause more than 80 per cent of all cases of MELAS. These mutations impair the ability of mitochondria to make proteins, use oxygen, and produce energy. Researchers have not determined how changes in mitochondrial DNA lead to the specific signs and symptoms of MELAS. They continue to investigate the effects of mitochondrial gene mutations in different tissues, particularly in the brain.

## Neuropathy, Ataxia, and Retinitis Pigmentosa

Mutations in one mitochondrial gene, MT-ATP6, have been found in people with neuropathy, ataxia, and retinitis pigmentosa (NARP). The MT-ATP6 gene provides instructions for making a protein that is essential for normal mitochondrial function. This protein forms one part (sub-unit) of an enzyme called ATP synthase. This enzyme, which is also known as Complex V, is responsible for the last step of oxidative phosphorylation, in which a molecule called adenosine diphosphate (ADP) is converted to ATP. Mutations in the MT-ATP6 gene alter the structure or function of ATP synthase, reducing the ability of mitochondria to make ATP. It is unclear how this disruption in mitochondrial energy production leads to muscle weakness, vision loss, and the other specific features of NARP.

## Non-syndromic Deafness

Mutations in two mitochondrial genes, MT-RNR1 and MT-TS1, are associated with non-syndromic deafness (hearing loss without related signs and symptoms affecting other parts of the body). These genes provide instructions for making types of RNA. The MT-RNR1 gene provides instructions for a specific type of ribosomal RNA called 12S RNA. A particular form of transfer RNA, designated as tRNASer(UCN), is formed from the MT-TS1 gene. Both of these RNA molecules help assemble amino acids into full-length, functioning proteins within mitochondria.

Mutations in the MT-RNR1 gene increase the risk of hearing loss, particularly in people who take antibiotic medications called aminoglycosides. These antibiotics are typically used to treat chronic bacterial infections such as tuberculosis. Aminoglycosides kill bacteria by binding to their ribosomal RNA and disrupting the bacteria's ability to make proteins. Genetic changes in the MT-RNR1 gene often make the 12S RNA in human cells look similar to bacterial ribosomal RNA. As a result, aminoglycosides can target the altered 12S RNA just as they target bacterial ribosomal RNA. The antibiotic easily binds to the abnormal 12S RNA, which impairs the ability of mitochondria to produce proteins needed for oxidative phosphorylation. Researchers believe that this unintended effect of aminoglycosides may reduce the amount of ATP produced in mitochondria, increase the production of harmful byproducts, and eventually cause the cell to self-destruct (undergo apoptosis).

Non-syndromic deafness also results from genetic changes in the MT-TS1 gene. Most of the mutations change a single building block (nucleotide) in the tRNASer(UCN) molecule. These mutations likely disrupt the normal production of the molecule or alter its structure. As a result, less tRNASer(UCN) is available to assemble proteins within mitochondria. These changes reduce the production of proteins needed for oxidative phosphorylation, which may impair the ability of mitochondria to make ATP.

Researchers have not determined why the effects of mutations in the MT-RNR1 and MT-TS1 genes are usually limited to cells in the inner ear that are essential for hearing. They believe that other genetic or environmental factors must play a role in the signs and symptoms associated with these mutations.

## Other Disorders

Inherited changes in mitochondrial DNA can cause problems with growth, development, and function of the body's systems. These mutations disrupt the mitochondria's ability to efficiently generate energy for the cell. Conditions caused by mutations in mitochondrial DNA often involve multiple organ systems. The effects of these conditions are most pronounced in organs and tissues that require a lot of energy (such as the heart, brain, and muscles). Although the health consequences of inherited mitochondrial DNA mutations vary widely, some frequently observed features include muscle weakness and wasting, problems with movement, diabetes, kidney failure, heart disease, loss of intellectual functions (dementia), hearing loss, and abnormalities involving the eyes and vision.

A build-up of non-inherited (somatic) mutations in mitochondrial DNA has been associated with an increased risk of certain age-related disorders such as heart disease, Alzheimer disease, and Parkinson disease. Additionally, research suggests that the progressive accumulation of these mutations over a person's lifetime may play a role in the normal process of aging.

# 13

# Gene Therapy and Genetic Testing

## Introduction

Gene therapy may be used for treating, or even curing, genetic and acquired diseases like cancer and AIDS by using normal genes to supplement or replace defective genes or to bolster a normal function such as immunity. It can be used to target somatic (i.e., body) or germ (i.e., egg and sperm) cells. In somatic gene therapy, the genome of the recipient is changed, but this change is not passed along to the next generation. In contrast, in germline gene therapy, the egg and sperm cells of the parents are changed for the purpose of passing on the changes to their offspring.

There are basically two ways of implementing a gene therapy treatment:

1. Ex vivo, which means "outside the body" — Cells from the patient's blood or bone marrow are removed and grown in the laboratory. They are then exposed to a virus carrying the desired gene. The virus enters the cells, and the desired gene becomes part of the DNA of the cells. The cells are allowed to grow in the laboratory before being returned to the patient by injection into a vein.

2. *In vivo*, which means "inside the body" – No cells are removed from the patient's body. Instead, vectors are used to deliver the desired gene to cells in the patient's body.

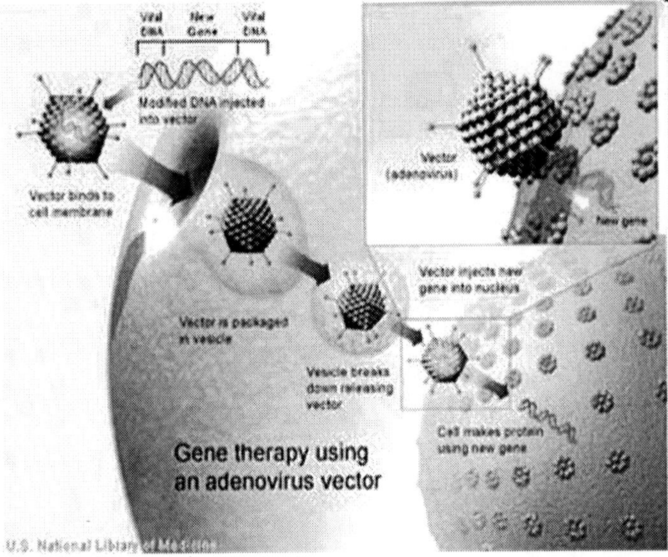


**Fig. 13.1:** Gene therapy using an Adenovirus vector. A new gene is inserted into an adenovirus vector, which is used to introduced the modified DNA into a human cell. If the treatment is successful, the new gene will make a functional protein.

Currently, the use of gene therapy is limited. Somatic gene therapy is primarily at the experimental stage. Germline therapy is the subject of much discussion but it is not being actively investigated in larger animals and human beings.

As of June 2001, more than 500 clinical gene-therapy trials involving about 3,500 patients have been identified worldwide. Around 78 per cent of these are in the United States, with Europe having 18 per cent. These trials focus on various types of cancer, although other multigenic diseases are being studied as well. Recently, two children born with severe combined immunodeficiency disorder ("SCID") were reported to have been cured after being given genetically engineered cells.

Gene therapy faces many obstacles before it can become a practical approach for treating disease. At least four of these obstacles are as follows:

1. Gene delivery tools. Genes are inserted into the body using gene carriers called vectors. The most common vectors now are viruses, which have evolved a way of encapsulating and delivering their genes to human cells in a pathogenic manner. Scientists manipulate the genome of the virus by removing the disease-causing genes and inserting the therapeutic genes. However, while viruses are effective, they can introduce problems like toxicity, immune and inflammatory responses, and gene control and targeting issues.
2. Limited knowledge of the functions of genes. Scientists currently know the functions of only a few genes. Hence, gene therapy can address only some genes that cause a particular disease. Worse, it is not known exactly whether genes have more than one function, which creates uncertainty as to whether replacing such genes is indeed desirable.
3. Multigene disorders and effect of environment. Most genetic disorders involve more than one gene. Moreover, most diseases involve the interaction of several genes and the environment. For example, many people with cancer not only inherit the disease gene for the disorder, but may have also failed to inherit specific tumour suppressor genes. Diet, exercise, smoking and other environmental factors may have also contributed to their disease.
4. High costs. Since gene therapy is relatively new and at an experimental stage, it is an expensive treatment to undertake. This explains why current studies are focussed on illnesses commonly found in developed countries, where more people can afford to pay for treatment. It may take decades before developing countries can take advantage of this technology.

## Human Genome Project

The Human Genome Project (HGP) was an international scientific research project with a primary goal to determine the sequence of chemical base pairs which make up DNA and to identify the approximately 25,000 genes of the human genome from both a physical and functional standpoint.

The project began in 1990 initially headed by James D. Watson at the U.S. National Institutes of Health. A working draft of the genome was released in 2000 and a complete one in 2003, with further analysis still being published. A parallel project was conducted by the private company Celera Genomics. Most of the sequencing was performed in universities and research centers from the United States, Canada and Britain. The mapping of human genes is an important step in the development of medicines and other aspects of health care.

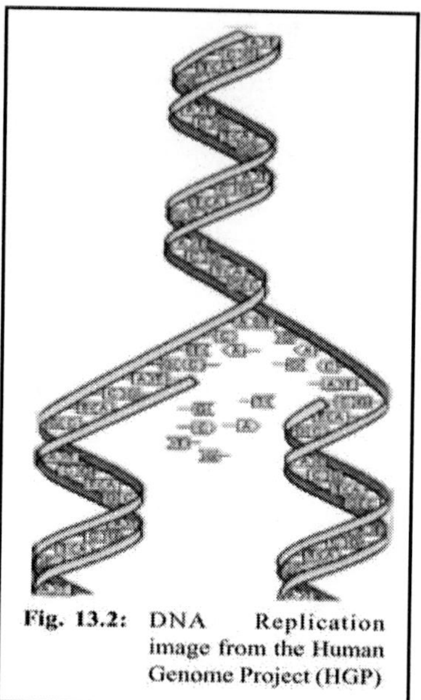

**Fig. 13.2:** DNA Replication image from the Human Genome Project (HGP)

While the objective of the Human Genome Project is to understand the genetic make-up of the human species, the project also has focussed on several other non-human organisms such as E. coli, the fruit fly, and the laboratory mouse. It remains one of the largest single investigational projects in modern science.

The HGP originally aimed to map the nucleotides contained in a haploid reference human genome (more than three billion). Several groups have announced efforts to extend this to diploid human genomes including the International HapMap Project, Applied Biosystems, Perlegen, Illumina, JCVI, Personal Genome Project, and Roche-454.

The "genome" of any given individual (except for identical twins and cloned animals) is unique; mapping "the human genome" involves sequencing multiple variations of each gene. The project did not study the entire DNA found in human cells; some heterochromatic areas (about 8% of the total) remain un-sequenced.

## The Project

Initiation of the Project was the culmination of several years of work supported by the United States Department of Energy, in particular workshops in 1984 and 1986 and a subsequent initiative of the US Department of Energy. This 1987 report stated boldly, "The ultimate goal of this initiative is to understand the human genome" and "knowledge of the human genome is as necessary to the continuing progress of medicine and other health sciences as knowledge of human anatomy has been for the present state of medicine." Candidate technologies were already being considered for the proposed undertaking at least as early as 1985.

James D. Watson was head of the National Center for Human Genome Research at the National Institutes of Health (NIH) in the United States starting from 1988. Largely due to his disagreement with his boss, Bernadine Healy, over the issue of patenting genes, he was forced to resign in 1992. He was replaced by Francis Collins in April 1993, and the name of the Center was changed to the National Human Genome Research Institute (NHGRI) in 1997.

The $3-billion project was formally founded in 1990 by the United States Department of Energy and the U.S. National Institutes of Health, and was expected to take 15 years. In addition to the United States, the international consortium comprised geneticists in China, France, Germany, Japan, and the United Kingdom.

Due to widespread international cooperation and advances in the field of genomics (especially in sequence analysis), as well as major advances in computing technology, a 'rough draft' of the genome was finished in 2000 (announced jointly by then US president Bill Clinton and British Prime Minister Tony Blair on June 26, 2000). Ongoing sequencing led to the announcement of the essentially complete genome in April 2003, 2 years earlier than planned. In May 2006, another milestone was passed on the way to completion of the project, when the sequence of the last chromosome was published in the journal Nature.

## State of Completion

There are multiple definitions of the "complete sequence of the human genome". According to some of these definitions, the genome has already been completely sequenced, and according to other definitions, the genome has yet to be completely sequenced. There have been multiple popular press articles reporting that the genome was "complete." The genome has been completely sequenced using the definition employed by the International Human Genome Project. A graphical history of the human genome project shows that most of the human genome was complete by the end of 2003. However, there are a number of regions of the human genome that can be considered unfinished:

- First, the central regions of each chromosome, known as centromeres, are highly repetitive DNA sequences that are difficult to sequence using current technology. The centromeres are millions (possibly tens of millions) of base pairs long, and for the most part these are entirely un-sequenced.
- Second, the ends of the chromosomes, called telomeres, are also highly repetitive, and for most of the 46 chromosome ends these too are incomplete. It is not known precisely how much sequence remains before the telomeres of each chromosome are reached, but as with the centromeres, current technological restraints are prohibitive.
- Third, there are several loci in each individual's genome that contain members of multigene families that are difficult to disentangle with shotgun sequencing methods - these multigene families often encode proteins important for immune functions.
- Other than these regions, there remain a few dozen gaps scattered around the genome, some of them rather large, but there is hope that all these will be closed in the next couple of years.

In summary: the best estimates of total genome size indicate that about 92 per cent of the genome has been completed and it is likely that the centromeres and telomeres will remain un-sequenced until new technology is developed that facilitates their sequencing. Most of the remaining DNA is highly repetitive and unlikely to contain genes, but it cannot be truly known until it is entirely sequenced. Understanding the functions of all the genes and their regulation is far from complete. The roles of junk DNA, the evolution of the genome, the differences between individuals, and many other questions are still the subject of intense interest by laboratories all over the world.

## Goals

The sequence of the human DNA is stored in databases available to anyone on the Internet. The U.S. National Center for Biotechnology Information (and sister organisations in Europe and Japan) house the gene sequence in a database known as Genbank, along with sequences of known and hypothetical genes and proteins. Other organisations such as the University of California, Santa Cruz, and Ensemblpresent additional data and annotation and powerful tools for visualising and searching it. Computer programmes have been developed to analyse the data, because the data themselves are difficult to interpret without such programmes.

The process of identifying the boundaries between genes and other features in raw DNA sequence is called genome annotation and is the domain of bioinformatics. While expert biologists make the best annotators, their work proceeds slowly, and computer programmes are increasingly used to meet the high-throughput demands of genome sequencing projects. The best current technologies for annotation make use of statistical models that take advantage of parallels between DNA sequences and human language, using concepts from computer science such as formal grammars.

Another, often overlooked, goal of the HGP is the study of its ethical, legal, and social implications. It is important to research these issues and find the most appropriate solutions before they become large dilemmas whose effect will manifest in the form of major political concerns.

All humans have unique gene sequences. Therefore the data published by the HGP does not represent the exact sequence of each and every individual's genome. It is the combined genome of a small number of anonymous donors. The HGP genome is a scaffold for future work in identifying differences among individuals. Most of the current effort in identifying differences among individuals involves single nucleotide polymorphisms and the HapMap.

Almost all the goals that the Human Genome Project has set for itself have been completed earlier than predicted. The Human Genome Project actually exceeded the projected finishing time by two years. The Human Genome Project set a reasonable, attainable goal of 95 per cent of DNA to be sequenced. Not only did the researchers surpass that goal, they shattered their prediction, and were able to sequence 99.99 per cent of a human's DNA. Not only did The Human Genome Project exceed all goals and standards, it still continues making progress on those goals already achieved.

## How It was Accomplished

Funding came from the US government through the National Institutes of Health in the United States, and the UK charity, the Wellcome Trust, who funded the Sanger Institute (then the Sanger Centre) in Great Britain, as well as numerous other groups from around the world. The genome was broken into smaller pieces; approximately 150,000 base pairs in length. These pieces were then spliced into a type of vector known as "bacterial artificial chromosomes", or BACs, which are derived from bacterial chromosomes which have been genetically engineered. The vectors containing the genes can be inserted into bacteria where they are copied by the bacterial DNA replication machinery. Each of these pieces was then sequenced separately as a small "shotgun" project and then assembled. The larger, 150,000 base pairs go together to create chromosomes. This is known as the "hierarchical shotgun" approach, because the genome is first broken into relatively large chunks, which are then mapped to chromosomes before being selected for sequencing.

## Genetic Testing

Genetic testing allows the genetic diagnosis of vulnerabilities to inherited diseases, and can also be used to determine a person's ancestry. Normally, every person carries two copies of every gene, one inherited from their mother, one inherited from their father. The human genome is believed to contain around 20,000-25,000 genes. In addition to studying chromosomes to the level of individual genes, genetic testing in a broader sense includes biochemical tests for the possible presence of genetic diseases, or mutant forms of genes associated with increased risk of developing genetic disorders. Genetic testing identifies changes in chromosomes, genes, or proteins. Most of the time, testing is used to find changes that are associated with inherited disorders. The results of a genetic test can confirm or rule out a suspected genetic condition or help determine a person's chance developing or passing on a genetic disorder. Several hundred genetic tests are currently in use, and more are being developed.

Since genetic testing may open up ethical or psychological problems, genetic testing is often accompanied by genetic counselling.

## Types

Genetic testing is "the analysis of RNA, chromosomes (DNA), proteins, and certain metabolites in order to detect heritable disease-related genotypes, mutations, phenotypes, or karyotypes for clinical purposes ". It can provide information about a person's genes and chromosomes throughout life. Available types of testing include:

- *Newborn screening:* Newborn screening is used just after birth to identify genetic disorders that can be treated early in life. The routine testing of infants for certain disorders is the most widespread use of genetic testing—millions of babies are tested each year in the United States. All states currently test infants for phenylketonuria (a genetic disorder that causes mental retardation if left untreated) and congenital hypothyroidism (a disorder of the thyroid gland).

- *Diagnostic testing:* Diagnostic testing is used to diagnose or rule out a specific genetic or chromosomal condition. In many cases, genetic testing is used to confirm a diagnosis when a particular condition is suspected based on physical mutations and symptoms. Diagnostic testing can be performed at any time during a person's life, but is not available for all genes or all genetic conditions. The results of a diagnostic test can influence a person's choices about health care and the management of the disease.
- *Carrier testing:* Carrier testing is used to identify people who carry one copy of a gene mutation that, when present in two copies, causes a genetic disorder. This type of testing is offered to individuals who have a family history of a genetic disorder and to people in ethnic groups with an increased risk of specific genetic conditions. If both parents are tested, the test can provide information about a couple's risk of having a child with a genetic condition.
- *Prenatal testing:* Prenatal testing is used to detect changes in a fetus's genes or chromosomes before birth. This type of testing is offered to couples with an increased risk of having a baby with a genetic or chromosomal disorder. In some cases, prenatal testing can lessen a couple's uncertainty or help them decide whether to abort the pregnancy. It cannot identify all possible inherited disorders and birth defects, however.

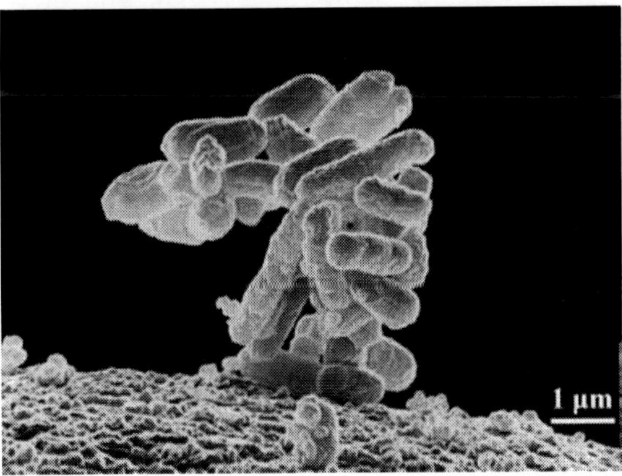


**Fig. 13.3:** The bacterium E. coli is routinely genetically engineered

- *Predictive and presymptomatic testing:* Predictive and presymptomatic types of testing are used to detect gene mutations associated with disorders that appear after birth, often later in life. These tests can be helpful to people who have a family member with a genetic disorder, but who have no features of the disorder themselves at the time of testing. Predictive testing can identify mutations that increase a person's chances of developing disorders with a genetic basis, such as certain types of cancer. For example, an individual with a mutation in BRCA1 has a 65 per cent cumulative risk of breast cancer. Presymptomatic testing can determine whether a person will develop a genetic disorder, such as hemochromatosis (an iron overload disorder), before any signs or symptoms appear. The results of predictive and presymptomatic testing can provide information about a person's risk of developing a specific disorder and help with making decisions about medical care.
- *Forensic testing:* Forensic testing uses DNA sequences to identify an individual for legal purposes. Unlike the tests described above, forensic testing is not used to detect gene mutations associated with disease. This type of testing can identify crime or catastrophe victims, rule out or implicate a crime suspect, or establish biological relationships between people (for example, paternity).
- *Research testing:* Research testing include finding unknown genes, learning how genes work and advancing our understanding of genetic conditions. The results of testing done as part of a research study are usually not available to patients or their healthcare providers.

## Controversial Questions

Several issues have been raised regarding the use of genetic testing:

1. *Absence of cure.* There is still a lack of effective treatment or preventive measures for many diseases and conditions now being diagnosed or predicted using gene tests. Thus, revealing information about risk of a future disease that has no existing cure presents an ethical dilemma for medical practitioners.

2. *Ownership and control of genetic information.* Who will own and control genetic information, or information about genes, gene products, or inherited characteristics derived from an individual or a group of people like indigenous communities? At the macro level, there is a possibility of a genetic divide, with developing countries that do not have access to medical applications of biotechnology being deprived of benefits accruing from products derived from genes obtained from their own people. Moreover, genetic information can pose a risk for minority population groups as it can lead to group stigmatisation.

   At the individual level, the absence of privacy and anti-discrimination legal protections in most countries can lead to discrimination in employment or insurance or other misuse of personal genetic information. This raises questions such as whether genetic privacy is different from medical privacy.

3. *Reproductive issues.* These include the use of genetic information in reproductive decision-making and the possibility of genetically altering reproductive cells that may be passed on to future generations. For example, germline therapy forever changes the genetic make up of an individual's descendants. Thus, any error in technology or judgment may have far-reaching consequences. Ethical issues like designer babies and human cloning have also given rise to controversies between and among scientists and bioethicists, especially in the light of past abuses with eugenics.

4. *Clinical issues.* These center on the capabilities and limitations of doctors and other health-service providers, people identified with genetic conditions, and the general public in dealing with genetic information.

5. *Effects on social institutions.* Genetic tests reveal information about individuals and their families. Thus, test results can affect the dynamics within social institutions, particularly the family.

6. Conceptual and philosophical implications regarding human responsibility, free will vis-à-vis genetic determinism, and the concepts of health and disease.

## Field Description

Cytogenetics is the specialised area of laboratory medicine involving the study of normal and abnormal chromosomes and their relationship to human development and disease. In medical practice, the study of human chromosomes is important because changes in the chromosome number and structure can lead to birth defects, mental retardation, infertility, miscarriage and cancer.

Analysis of chromosomes can aid in the diagnosis, prognosis and monitoring of treatment involving conditions seen by medical geneticists, paediatricians, obstetricians, gynaecologists, perinatologists, hematologists, oncologists, endocrinologists, pathologists, urologists, internists and family practice physicians.

Characteristics of a typical cytogenetic technologist:

- enjoys working independently
- applies meticulous attention to detail
- thrives with high degree of responsibility
- likes working with visual stimulus

## More About Cytogenetics

Cytogenetic technologists analyze chromosomes using tissue cultures and slide preparations from peripheral blood, bone marrow, amniotic fluid, products of conception and tumour samples.They makemicroscopic evaluations of the chromosome number and morphology, and prepare reports of the findings for physicians. Cytogenetic technologists use fluorescent-labelled DNA to detect gene and chromosome abnormalities associated with birth defects and cancers.This technique, fluorescence *in situ* hybridisation (FISH), has become the most rapidly growing area in cytogenetics.

Cytogenetic technologists interact with cytogeneticists, clinical geneticists, and genetic counsellors to provide valuable information to health care providers from all aspects of medicine.

Cytogenetic laboratory personnel need to know about specimen requirements, handling and culturing living cells, chromosome morphology, chromosome abnormalities, methods of chromosome analysis and interpreting results of cytogenetic studies.

Clinical cytogenetics began in 1956 when the normal number of human chromosomes in each cell was established. Since then, many syndromes and diseases have been correlated with chromosome changes.

Molecular cytogenetics began in the late 1980s, when fluorescent-labelled nucleic acid probes were first hybridised with chromosomes. These FISH probes can be detected on metaphase chromosomes, in interphase nuclei, in tissue sections, or even in blastomeres or gametes. The FISH probes can hybridise to entire chromosomes or to single unique DNA sequences.

The type of FISH probe used is based on the type of syndrome/disease/condition that is suspected. The applications of FISH include ploidy analysis, translocation and structural breakpoint analysis, detection of deletions and duplications too small to be observed by conventional cytogenetics (microdeletions/microduplications), and gene mapping. FISH is often used as a powerful adjunct to conventional cytogenetics.

New techniques allow for increased resolution of chromosome banding patterns, permitting differentiation of a greater number of abnormalities. More than 20 different techniques are used for visualising chromosomes. Chromosome abnormalities are classified as numeric or structural anomalies. The anomalies can involve one, two or multiple chromosomes. Specific cytogenetic nomenclature has been developed to describe the cytogenetic findings based on chromosome number, sex chromosome complement, and presence or absence of chromosome abnormalities.

## Career Opportunities

Career opportunities for cytogenetic technologists are excellent, as cytogenetic technology remains a growing field. Clinical laboratory technologists and technicians held about 295,000 jobs in 2000. About half worked in hospitals. Cytogenetic technologists also are employed in clinical laboratories, research laboratories and cytogenetic-related biotechnology companies.

Cytogenetics is a dynamic field, both in technological developments and clinical applications. Moreover, the current needs for cytogenetic technologists exceed the supply of trained staff. Over the past several years, approximately 20 technologists have been hired each year by the Mayo Clinic Cytogenetic Laboratory. Our laboratory staff find their jobs interesting and rewarding.

If you enjoy independent, meticulous, microscopic work, and are comfortable with a high degree of responsibility, cytogenetics can provide great career satisfaction while you serve in a vital health care role.

# Index

❑❑❑